T0231782

From Atoms to Higgs Boson

Voyages in Quasi-Spacetime

Chary Rangacharyulu
Christopher Polachic

JENNY STANFORD
PUBLISHING

Published by

Jenny Stanford Publishing Pte. Ltd.
Level 34, Centennial Tower
3 Temasek Avenue
Singapore 039190

Email: editorial@jennystanford.com
Web: www.jennystanford.com

British Library Cataloguing-in-Publication Data
A catalogue record for this book is available from the British Library.

From Atoms to Higgs Boson: Voyages in Quasi-Spacetime

ISBN 978-981-4800-24-4 (Hardcover)
ISBN 978-0-429-02765-9 (eBook)

*This book is dedicated
to those men and women,
in all times, young in heart and soul,
who devote themselves to fathom
the submicroscopic world
and unravel its mysteries.*

Contents

Preface

This book has taken a few years of careful thinking, from start to finish, although its roots go back much further than that. It has grown out of the reality of how we teach and learn physics in higher education.

Every physics professor, especially a young one, faces the challenges of being an active, productive researcher in knowledge creation and an educator in knowledge transmission to the next generation. They return each semester to the challenge of presenting their undergraduate students with the set of standard topics prescribed in their syllabi and chosen textbooks. The flow of topics must cleave to an orthodoxy and reach its conclusion within the strict deadlines of the academic calendar.

The students also face their own demands in terms of time management and finding space in their brains to accommodate a growing knowledge of physics alongside information from many other, different courses. They, too, have their deadlines as they struggle to complete labs and assignments on time and adequately prepare themselves for graduate work or a successful career.

For both parties, success in these time-sensitive obligations requires a commitment to pragmatism. Although many undergraduate students would probably embrace the opportunity to contemplate the deeper philosophical aspects of physics and the structure of nature with their professors, the latter do not have the luxury of time or preparation to guide students in confronting some of the most interesting questions of this kind. The exception to this would be, perhaps, a more rigorous examination of the mathematical formalism and empirical evidence pertaining to their standard curricular topics, but these discussions do not necessarily probe the implicit mathematical and physical assumptions and philosophical scaffolding in sufficient detail.

Textbooks offer little help in this regard. They present the subject material as well-established facts suggesting that students (and instructors) who fail to unhesitatingly accept the subject matter should doubt their own intelligence. The textbook presentations thus encourage a perspective that physics has developed its orthodoxy along the lines of historical inevitability: If Einstein had not devised his special theory of relativity, including its metaphysical commitments, it was only a matter of time before someone else arrived at precisely the same conclusions and nuances. It must be so because all the content of our syllabus describes nature as it is, chapter and verse, takes note of it, and writes it down.

In this pedagogical tradition, it is not uncommon that a professor enthusiastically asserts, without reference to the awkward contextual background information, that in their state of special genius the founding fathers of modern physics (especially Einstein and Bohr) brought forth revolutionary ideas as mere postulates. The more counterintuitive the revolution, the better, because this enhances the mystery of the professional physicist as a gatekeeper and inheritor of a spark of this genius. This is quite frustrating to a student when a professor is pontificating on the self-evident truths of particle–wave duality, four-dimensional spacetime, or various other mathematical constructs inaccessible to direct experimental verification. Even if a professor has time for contemplation of these matters and how they might be best introduced and justified to students (which is very doubtful), years of devotion is required to weed through the vast amounts of not-so-easy-to-access literature of the nineteenth and early twentieth centuries (much of it in German) and assess all the false starts, varying views, doubts, or limits of usefulness of theoretical arguments expressed by the proponents of these "revolutionary concepts," prior to their ultimate acceptance as dogma. Thus, despite the best intentions of any physics professor, it is difficult to teach a subject as profound as modern physics while also understanding—let alone communicating to students—the real, historical, contextual limitations of our supposed knowledge of physical reality.

The authors of this book have traveled this voyage, this quest for the knowledge of physical reality, together for some years: first as a professor–student collaboration and later as two researchers and educators (one old and one young professor), working together with the same common interest. Both of us believe that we are not here to hand the knowledge of true physics down as established facts to aspiring young minds, but to engage these minds, insofar as this is possible within a tight university curriculum setting. This book is the product of an ongoing search comprising nearly 50 man-years between the two of us.

Over this time, we have seen that physics, as a subject of academic inquiry, has two faces. First, it is application oriented: The discipline has ever been, and continues to be, an enterprise that yields extraordinary results in utilitarian, technological innovation resulting in human flourishing. Second, physics sets itself to uncover and unravel the conceptual foundations of the dynamics of our physical universe, motivated by the innate curiosity of our species to simply understand what is true and real.

In the first aspect of our discipline, operational definitions, effectively expressed in mathematical symbols, are quite useful without necessarily being grounded in physical reality. Unfortunately, these operational definitions and their associated mathematical formalism are also indiscriminately applied to the second task, relating to conceptual foundations. At this point, an assumption can often be made, out of habit, to ascribe to every detail of our useful formalism a correspondence to some element of physical reality. This may be justified during an initial stage in creating our mathematical models. However, as the usefulness of the models increases for the purpose of "solving" complex problems and thus advancing our goal of application, so also do the mathematical elements of the model tend to drift far away from what we originally understood them to represent.

To this state of affairs, we have given the name *quasirealism*, based on the *quasiparticle* approach of solving otherwise intractable problems, which was first employed in the field of condensed matter physics and then extended to nuclear physics and, ultimately, to the realm of theoretical particle physics. The following pages are our attempt to ask some healthy questions about the quasirealist philosophy in modern physics, through various illustrations and perhaps not a few bold—even controversial—postulates of our own. We do not pretend, however, that they are the work of genius, just honest concern. And we will let the readers decide for themselves. If nothing else, may the reading of these questions and ideas provide as much stimulation and enjoyment as we have had working them out.

As co-authors we both share the desire to thank our families for their cheerfulness during the process of drafting this manuscript. They have patiently supported us in many ways and accompanied us both physically and emotionally through the long stages of writing.

During the course of our careers in physics, we have worked with and encountered various types of physicists. Some of them are dedicated to probing the depths of the physical mysteries they encounter in their work. Others are techno-savvy, motivated by the challenge of working with hardware, software, and analytical methods. Still others are tactful managers who get things done. We have also seen eager young minds come to our offices with hopes to become the next Einstein; and non-science majors, having heard rumors that reality is stranger than fiction, arrive with a willingness to trust any claims we make about the bizarre and spooky world of relativity, quantum theory, and the subatomic world. We have had many fruitful discussions on physics with all these kinds of colleagues. The SPIE San Diego series of conferences on "The Nature of Light: What Are Photons?" also kept us focused on this subject for more than 14 years. We thank Professor Chandrasekhar Roychoudhuri for engaging the first author in this series.

We are very thankful to Kaylyn Olshanoski who created several important figures for this book. Also her constructive feedback from the perspective of an engineering physics undergraduate student was quite helpful. Thank you to Kiyoko Kato, as well, for the exceptional image used in our cover design. It visually captures the intricacy and beauty of our subject matter and is worth glancing back at every-so-often as you peruse the book.

We are particularly grateful to our publisher, Stanford Chong, for suggesting that a book of this kind is desired and waiting patiently for it to be finally produced with extended deadlines. We also thank Jenny Rompas and Shivani Sharma of Jenny Stanford Publishing for their indispensable assistance in formatting and proofing our text and providing helpful guidance as we worked through the process of publication.

Chary Rangacharyulu
Christopher Polachic
Spring 2019

Introduction

The French don't care what they do, actually, as long as they pronounce it properly.

—Professor Henry Higgins
My Fair Lady (1964)

Our task is to learn to use these words correctly—that is, unambiguously and consistently.

—Niels Bohr[1]

Physics had its beginnings in the work of ancient natural philosophers. These experts esteemed the power of human reason, alone, as the chief means to discern truths about the fundamental structure and operations of nature—in a sense, through careful attention to the pronunciation of ideas and the correct and consistent use of words referring to features of the physical world, whether visible or invisible. In modern times, physics has assumed a role that is somewhat different from these beginnings. It is hailed as the most fundamental, rigorous, and hardcore of the sciences. Its practitioners are ultimately recognized for excellence in technological innovation and unrelenting progress in taming the secrets of the physical universe through objective experimental methods. At the heart of every technological innovation—whether computers, health instrumentation, agricultural technology, energy production, communications, or any of the myriad other possible categories—physics models are operational. It is inarguably impressive that physicists, using mathematical modeling based on a few basic conservation laws and associated symmetries, have been able to probe the interiors of the minutest of minute entities and to

---

[1]As quoted by Jørgen Kalckar, "Niels Bohr and his youngest disciples," Ed. S. Rozental, *Niels Bohr—His Life and Work as Seen by His Friends and Colleagues* (John Wiley: New York, 1967), 227–238.

realize applications that were hardly imaginable just a few decades ago. Seen in this light, the most obvious success of the discipline of physics cannot be measured by words pronounced and used correctly (as Professors Higgins and Bohr suggest), but by tangible achievements in the form of new technology.

The practice of physics involves an extensive study of complex systems composed of denumerably infinite degrees of freedom. Physicists reduce these unwieldy systems to computationally tractable equivalents. Inevitably, over the long history of this activity, physicists have taken the liberty to develop their own vocabulary, often derived from our day-to-day language rooted in common-sense perceptions, and from that of cognate disciplines. Physicists tend to communicate among themselves using familiar words and phrases endowed with novel meanings and significance, referring to abstract elements of the effective[2] theories they use to describe these complex systems. There are even differences in how the same term is used from one sub-discipline of physics to another. For example, in condensed matter physics, which focuses on the science of solid and liquid materials, an "electron" is not necessarily the same entity whose nominal mass we find in the data tables used by particle physicists, nor the one we refer to when we teach undergraduate physics courses. For plasma physicists, a "photon" is not always something that has zero mass, traveling at the constant speed of light, as it does for other physicists.

These discrepancies in vocabulary arise because of a tendency to retain the original labels when physicists develop and employ effective models of physical reality. These models are powerful and

---

[2]The word *effective* is an example of this very point. Common usage of this term implies the notion of "appropriateness" and "accomplishing the intended result." Thus, an *effective theory* might be understood outside of physical science circles to mean a scientific theory that appropriately captures the real physical attributes of a natural system: It is *effective* in doing what it is supposed to do. To the physicist, however, *effective* most often denotes a technical simplification that replaces real parameters of a tremendously complicated system with statistical averages or abstract substitutions. These retain important similarities to the physical system of interest but no longer provide direct information about the physical world. To a physicist, then, an *effective theory* allows progress to be made in finding *some* solution to a problem, if not the one we really desire. The theory must be carefully interpreted to discern what information this progress yields about the original physical system of interest. Effective theories and models are, of course, common in other disciplines as well, such as economics.

useful in large part because they can retain a logical connection back to the original physical concepts, but the connection must be carefully followed, like a trail of breadcrumbs leading one back through a tangled forest. It is thus natural to re-use familiar words for the new, effective entities or concepts. However, taken too literally, this re-use of common labels may lead to surprising results. In the leading effective model of superconductivity, two interacting "electron" partners, called a *Cooper pair*, are not located at any particular position in a superconducting medium but are present throughout the material at any given time. Physicists can mathematically define lasers as negative temperature devices (below zero Kelvin). They can define entities in their experiments that behave like material *particles* having negative mass, which accelerate in a direction opposite to an applied force.

Properly understood, none of these is a deliberate equivocation or misuse of terminology, so long as we recall the essential deviation we have taken from reality when we devise a solvable effective theory. Without that care, however, the conceptual consequences that arise may be the source of endless astonishment among non-specialists, providing fodder for exciting headlines in popular science news reporting. Negative mass! Faster-than-light speeds! Curved spacetime! Rather than viewing these novelties with consternation, the public trusts that the details they glean, leaking out of physicists' effective models, must be referring to something fundamentally real in our physical world: a revelation that would be otherwise inaccessible to humanity, if not for the inspired proclamations of the scientific magisterium. This trust should not be dismissed as naïve: The greatest *apologia* for modern physics is technological application. Technology talks.

Physics has a broad scope. This is, in fact, an outrageous understatement. Over the centuries, the discipline has addressed questions about the physical universe at astronomical scales as well the tiniest entities comprising the heart of matter and has offered descriptions of how nature operates in the present as well as what its properties and rules were at the beginning of the universe. Physicists have even weighed in as eschatologists, employing current models to speculate on the future evolution and ultimate end of the cosmos. The inclination to apply physical theories to a description of the universe as a whole was, in an earlier time in the Christian and

Muslim West, a useful "handmaiden" to theological studies, confined to the investigations of *natural theology* in which sensory data from the Book of Nature could provide insight into interpretations of revealed knowledge from the Book of Scripture.

Over time, the study of the cosmos took on a life of its own in physics, disengaging from its theological context. It has now become an overarching narrative that motivates the entire industry of fundamental research. The unifying theme of this secular cosmology is no longer supernatural metaphysics, but the vision of physical reductionism. That is to say, if we can uncover the ultimate, basic building blocks of the universe (however tiny and practically inaccessible they may be) and determine how they interact with one another at all energies and at all distances (however close or far apart they are), then we can tame the universe within the mathematical prescription of some ultimate, all-encompassing model of physical reality in its entirety.

Fortunately for physicists, at least as a first approximation, energies and distances are not uncorrelated. It is understood that any useful probe of the most miniscule interior structures of matter will require experiments involving higher energies. In the first half of the twentieth century, physicists hurled electrons, protons, or neutrons onto protons in an attempt to study the interactions that occur between bits of matter at short distances. Surprisingly, as energies increased, hitherto unknown new entities made their appearance as transient bodies borne out of the explosive energies in these collisions. What was expected to be a simple measurement of two small bodies scattering off one another became a complex analysis of many bodies possessing new information about physical reality.

Undeterred by this development—indeed, emboldened by the opportunity to investigate "new physics"—physicists have marched on. The post-Second World War era saw the rise of particle physics as a distinct area of investigation, equipped with a generous kit of theoretical and experimental tools developed over many decades to probe the mysteries of the subatomic world. Particle physics now makes its own reductionist contribution to our knowledge of matter, and thus to a grand theory of the structure and history of the entire universe. Today, this toolkit takes the form of complicated, effective mathematical models and immense technological devices designed to access the behavior of bits of matter at enormous energy scales. Through familiarity with these effective models, the language of particle physics has taken on a life of its own. The word "particle," for example, and the concept of "mass" have subtly but profoundly diverged from the way these terms were understood only a century ago, and the way most people outside of the discipline—including other scientists—still think of them today.

This book is an effort to understand and humbly critique the accomplishments, logical rigor, and faithfulness to basic notions about physical reality that have accompanied the modern quest to unravel the physics of the universe. Of particular importance to our analysis—indeed, the central formal point behind this work—is the influence of effective models and theories on the way physicists think and talk about physical reality. This book is not meant to call into question any of the elegance, technical competence, or ingenious problem-solving that continues to define the outstanding work done by the international community of physicists who are investigating the world in which we live. We only hope to raise a serious concern about interpretation of results. It is a concern of metaphysics, really, and of ensuring that what the community of physicists says about our results to the listening world (and to ourselves and our students) is *true*, and not just *useful*. Mathematical models can be extraordinarily useful, without directly providing true knowledge about physical reality. This distinction is what we want our colleagues and those who are non-professionally interested in physics to consider more carefully.

The initial two chapters, together, define our larger concern with the metaphysical confusion just explained. The first chapter sets the stage with a discussion of reductionism as the driving motive of modern physics. In the second chapter, we devise and illustrate a new term: *quasirealism*. Quasirealism involves a failure to distinguish effective theoretical concepts from real, physical entities and properties. It is a term we use throughout the book. Any reader more comfortable with physics apart from philosophy may prefer to skim these chapters on first reading, becoming familiar with the main ideas, and return later on to fill in the details.

One can think of each subsequent chapter as offering a variation on the central theme that quasirealist assumptions dramatically affect the conclusions that physicists draw about the fundamental nature of the world in which we live. Each of these chapters focuses on a different concept in the current of modern physics, and while they are ordered to build conceptually upon one another, they may be read in isolation and out of order without too much detriment to our overall point (we hope). The flow, however, begins with big ideas and moves toward specificity by the end of the book.

The earlier chapters of this book involve a discussion of mass, space, and time, concepts that form the background of all discussions in physics. For physicists, the meaning of these two words has changed dramatically in the past 200 years, and this is scarcely acknowledged except among the most careful treatments by philosophers and interested scientists. Already in the late nineteenth century it was realized that a thorough understanding of space and time, in which entities exist and interactions occur, is essential for a proper description of physical dynamics. In a formal sense, the study of material objects in space requires a background assumption of geometrical relations that were provided with great clarity and objective physical consistency by the Greek mathematician Euclid. Over nearly two millennia the quest to improve upon and advance beyond Euclidean geometrical postulates led to non-Euclidean geometries. Although not obviously related to the properties of our own spatial reality, these new geometries yielded important mathematical developments.[3] It was only a matter of time before they were adopted into a serious physical framework: Einstein's theories of relativity.

This leads quickly and naturally to a discussion of mathematical spaces, such as Hilbert space and complex spaces, and we argue that physicists should take special care to avoid the confusion (both of themselves as well as their aspiring students and the general public) that may arise if conceptual, metaphysical rigor is lost in favor of mathematical gain. The influence of quasirealism cannot be discounted in all these conceptual maneuvers, and we attempt to raise important questions about this influence and how it leaves us impoverished.

We continue our discussion with a move to the smallest scales, examining the development of quantum theory and its connection to quasirealist interpretation. In that light, we explore the meaning of

---

[3]We hasten to add that the terms "geometry" and "space" are often erroneously used as synonyms.

elementary quanta, considered to be fundamental pieces of physical reality. This involves us with the classical concept of *atomicity*, which until recently[4] centered on the notion of physical indivisibility. Our discussion explores how the modern concept of atomicity and the word *particle* have evolved with reference to the building blocks of ordinary matter and energy, identified by particle physicists as electrons, quarks, and photons. In the case of quarks, the question of interest is how much ontological reality to allow these apparent entities. To what extent can we consider them physically real when they are defined in our leading theories primarily as composite mathematical structures, built on deeper mathematical forms, only loosely related to anything tangible that one might observe in a well-designed experiment?

We separately discuss electrons and photons, raising the concern that quasirealist interpretations have confused our picture of these particles in the modern physics framework, taking us further away from the reductionist aims of physics. In the case of the electron, it was the first fundamental part of subatomic matter to be observed directly as a simple corpuscle, confirming Dalton's (and Democritus') atomic hypothesis. In that discussion we return to the topic of quantum theory to show, in the specific example of the electron, how quasirealist quantum mechanical interpretations served to confound physicists' thinking about the physical properties of this otherwise uncontroversial species of the subatomic world.

All of this necessitates some discussion of the quantum theory of fields, wherein photons have friends among the other gauge bosons, namely the W, Z, and gluons (besides the gravitons). We examine the role of these bosons in particle physics models, and the assignment of their properties as suggested by theory and constrained by experiment. The thinking behind this book was inspired by the announcement of the Higgs boson's discovery at CERN—a momentous event in the history of physics—and so it is natural that we devote an entire concluding chapter to this entity and the determination of its unique properties. After all, it has been delegated the angelic assignment of providing mass to all other bodies in the universe.

As already said, the reader should find that the central theme remains basically the same throughout the parts of this book,

---

[4]"Recently" with respect to the grand sweep of history, at least.

restated with different emphases: Quasirealist perspectives have clouded the noble reductionist vision of modern physics, and the conclusions we think we are reaching about physical reality—especially those derived from particle physics—may be moving us in the wrong direction from where we think we are going. Even if this concern is found to be misguided, we hope that the analysis we provide in the following pages will serve to generate conversation among the most important audience this book can reach: the next generation of aspiring physicists, studying in their discipline as undergraduate and graduate students, dreaming the same dreams as their illustrious predecessors that they might someday, soon, help to expand our shared knowledge of the underlying, real stuff that makes the world we live in what it is.

Chapter 1

The Reductionist Vision of Physics

Physics is, hopefully, simple. Physicists are not.

—Edward Teller
Conversations on the Dark Secrets of Physics (1991)

1.1 Reductionism

Let us begin with a bold but not very controversial claim: Reductionism is, and has always been, the scientist's guide for investigating and understanding the physical world. This is the case because the reductionist vision is simply an instinct of our common sense. In unvarnished terms, reductionism is a metaphysical view that the behavior of a composite system can be fully understood from the combined behavior of its constituent parts and nothing more.

In physics, a reductionist view of nature can be applied in two ways, not unrelated to one another. First, physicists may seek a unified, underlying theoretical description of all natural phenomena, reducing higher-order rules and theories to a single, grand unified causal theory with (in principle) universal application. We might refer to this as *theoretical* reductionism. The other reductionist strategy of physical science, which we can call *physical* or *ontological* reductionism, has to do with the physical structure of matter.

Physical reductionists understand matter to be made of a hierarchical arrangement of increasingly small components,

From Atoms to Higgs Boson: Voyages in Quasi-Spacetime
Chary Rangacharyulu and Christopher Polachic
Copyright © 2019 Jenny Stanford Publishing Pte. Ltd.
ISBN 978-981-4800-24-4 (Hardcover), 978-0-429-02765-9 (eBook)
www.jennystanford.com

ending—perhaps—at some fundamental microscopic level. The two reductionist ambitions of physics intersect in that the behavior of the final building blocks may be governed by a single, consistent, underlying physical theory, perhaps in the form of a small number of elegant mathematical equations.

Before we go on, we should briefly acknowledge that there are alternatives to reductionist ways of understanding natural systems. In particular, the perspective of *emergence* asserts that complex systems sometimes give rise to phenomena that (somehow) become ontologically independent of their underlying cause, in the sense that these emergent properties cannot, even in principle, be *fully* explained in terms of the mechanical operations of a more basic underlying order. Along with this loss of physical dependency, emergence naturally precludes the possibility of the unifying framework of theoretical reductionism.

For those who have trained in the reductionist school of physics, emergence is a difficult view to embrace. Giving the problem careful consideration, the physicist may acknowledge that some macroscopic properties of physical systems *appear* to be ontologically disconnected from underlying mechanisms and can be described by macroscopic terms and rules that have no immediate reference to microscopic reality,[1] but it cannot be admitted that there is *really* such an ontological or theoretical disconnect between scales of matter.

The reductionist vision is the framework that has successfully and usefully guided the arrow of scientific discovery since antiquity. *Physical* reductionism, in particular, is a powerful, intuitive idea and arguably a simple matter of common sense.

1.2 Our View of the World

Human beings are sensory beings. Our everyday sense perceptions provide us with data about the physical world. They are the natural starting point for a scientific description of nature.

The data obtained from our senses are all bound to a certain scale of physical reality: the *macroscopic* world of everyday-sized objects moving slowly enough that we can take notice and perform measurements. At this scale, our sensory data inform our

---

[1]For example, the macroscopic property of temperature is a net result of the kinematical and dynamical properties of the microscopic constituents in bulk matter.

reasoning about nature along three fundamental physical lines: spatial extension, time, and interaction. We observe a physical world composed of spatially extended objects that interact with one another by various means—usually, but not always, through physical contact—and which experience changes in their relative position and composition according to a certain ordering, which is congruent with a subjective sense of time passing. Our reductionist theories of physics have been necessarily expressed in these terms, because these are the basic categories that appear to define the world in which we live.

Our natural observation of macroscopic physical systems makes it evident to us that material objects have composite structure. This is clear from the fact that apparently continuous material substances can be divided into smaller pieces. A volume of liquid can be separated into an arbitrarily large number of containers, and a large solid object can be broken by force into smaller components, each of which can be further reduced. Air around us is separated by the passage of a body moving through it. Even though our senses are unable to perceive objects smaller than a certain size, this pattern of reducible composition at larger scales suggests our everyday thinking that even below the limit of perception, this pattern should go on further.

At each scale of matter, these fragments should have the same properties of (progressively reduced) spatial extension, location, and motion relative to other bodies, all evidently describable in terms of the conservation of certain parameters of the system such as energy and momentum, and identifiable rules of interaction among the physical entities. Reducibility is simply an observed fact of macroscopic objects. The inference that reduction continues into microscopic scales of the physical world may not be strictly logical (in the sense of provable), but it is a natural extrapolation in the absence of contrary evidence.[2]

Science has been divided into disciplines that operate at these different scales. The biological scale of life can, in part, be understood reductionistically in terms of structural elements called

---

[2]This extrapolation may simply reflect a dearth of imagination on our part, and the principle of physical reduction gives up at some point below the limit of perception. This is not, however, what happens with quantum theory, which represents a definite change in the rules at such a scale and does not reflect a deviation from either physical or theoretical reductionism.

cells. Types of cells differ in form and function but share some common characteristics and have tremendous internal complexity. They are reducible to biochemical structures constituted by organic molecules of various sizes, compositions, and functions. But the incredible, dynamic variety of the molecular world is itself reducible to a relatively compact set of elemental atoms categorized in the periodic table of elements. If this view of the physical world is correct (and physics operates on the assumption that it is), then (in principle) if we could[3] fully describe the behavior of the atoms, we could work our way up to a complete physical description of the cells.

The reduction of molecules to the atomic scale is a significant step in simplification and organization of our knowledge of matter. About 100 chemical atomic species form the underlying structure of our visible, everyday world. The properties of these atomic constituents can be described by a conceptually simpler set of more fundamental building blocks. The essential structure of every atom is the combination of mutually interacting protons, neutrons, and electrons—three species of bricks forming the larger whole. The protons and neutrons are held together in the nucleus by the strong nuclear interaction, and the electrons bind to the protons via electromagnetic attraction. Three kinds of particles and two interactions combine in various stable configurations to build up the larger world in all its complexity, as individual atoms interact with one another through their electrons. The result is a physical, macroscopic world of gases, liquids, and solids. Understand the properties of the former scale, and you should, in principle, understand the larger. This is the power of reductionist vision.

The relatively small number of entities—three!—involved in the structure of atoms irresistibly suggests that the remaining journey to the fundamental level of matter may be short. The enterprise of nuclear physics has pursued this dream since the discovery of the

---

[3]We cannot, of course. Real natural systems are far too complex for an adequate theoretical description with realistic predictive power. But we are talking here about how nature works, with or without our ability to model it. The dynamic interplay between entities and their interactions at the microscopic level appears to be an insurmountable barrier to a complete theoretical reductionist description. This is a limitation of our epistemological resources, however, not a clue that nature is non-reductionist in structure.

neutron in 1932. Scientists have employed high-energy collisions of atomic constituents to try to break apart the nuclei. These efforts, rather than resulting in clear evidence of a lower level, instead produced an astonishing variety of exotic, short-lived particle-like states with properties similar to those of protons and neutrons.

Attempting to classify these new species, particle physicists settled on a reductionist model of quarks to describe the underlying structure of many of these new discoveries in the subatomic world. The theory of this quark model suggests a new, more basic level for the physical world below the triplicate foundation of electrons, protons, and neutrons. Do quarks provide a final answer to an ancient metaphysical question: "What is the bottom level to physical reductionism?" As we will argue in detail in this book, the answer should not be an unqualified "yes" because it is not at all clear that the quark model has continued along, rather than deviated from, the trajectory of the reductionist enterprise at the heart of physics.

1.3 Democritus' Atoms

The hunt for the foundational stuff of matter, both philosophically and experimentally, is an old one. Thales of Miletus opined that all substances in nature could be understood as different, complex forms of an underlying fundamental substance, namely water. Leucippus of Thracian Abdera, in the fifth-century BCE, mused whether the apparent continuity of the Aegean water was merely an illusion of scale, the substance being like the sand of the seashore, divisible.

But how far divisible? If a volume of water be divided into parts, can those parts be divided again, and then further divided, *ad infinitum*? Here, Leucippus' disciple Democritus concluded that common sense must prevail: There must be some smallest level of division at which the process stops—a fundamental unit with non-separable parts, which Democritus characterized as *uncuttable*, or *a-tomon* in Greek, from which our word *atom* is derived.

Democritus was a fifth-century BCE philosopher, also from Abdera. He was among the Greek *natural philosophers* who carefully applied their powers of reason to understand the rules and structures of nature, long before an experimental method had developed in the advent of modern science. None of Democritus' three dozen or so books on the natural sciences, mathematics, and medicine have survived down to today, except as fragments. However, his basic ideas were reiterated by later authors whose works do survive. Within the large body of quotations and commentary in the writings of Epicurus, Simplicius, and Aristotle, we have today a good sense of Democritus' hypotheses. For a man whose original writings no longer survive, he has attained a reputation as the archetype for ontological reductionism.

The basic stuff of matter, Democritus postulated, can be understood as eternally existing, small bodies, infinite in number. These atoms occupy empty space, *the void*, which has two characteristics: its emptiness and its infinite extent. The defining property of atoms is their indivisibility, which follows necessarily from reason: If there exists some substance of finite mass that is divisible into smaller parts at every point within its mass, then if all possible divisions were made it would be divided into points without spatial dimension. How much of the total mass would each point then contain? Can the recombination of points with no magnitude produce a larger macroscopic object of definite extension and finite mass? Democritus' answer was "no;" rather, the ultimate composition of matter must have some finite form and size and cannot be forever reducible. There is a necessary bottom layer to physical reality: the Democritian atoms.

It is important to notice that these atoms, while fundamental and microscopic, are not necessarily simple: Indivisibility does not require simplicity of structure. Democritus understood that the rich variety of complex structures in the macroscopic world requires some variety of form among the atoms, and thus different atomic species. Atoms, he thought, must have spatial extension—they are not points in space—and their different shapes and sizes allow interaction with one another without loss of individuality. They entangle and grasp one another, forming larger structures that can be broken apart by the application of a sufficient force to undo their bonds. This diversity of atomic species yields all the complex behaviors and bulk properties of matter. From the perspective of a modern high school or college education in the natural sciences, this ancient description of matter is beginning to sound very familiar.[4]

Democritus' physical reductionism offers no compromise for emergent phenomena: "The atoms," wrote Aristotle, interpreting Democritus' view,

> struggle and are carried about in the void because of their dissimilarities and the other differences mentioned, and as they are carried about they collide and are bound together in a binding which makes them touch and be contiguous with one another but which does not genuinely produce any other single nature whatever from them; for it is utterly silly to think that two or more things could ever become one.[5]

An emergent property—some quality that arises out of complexity in a non-reducible way—would be a signal of ontological unity coming from multiplicity, and this goes beyond the common-sense reductionism of the classical atomic theory.

While Democritus talked about atoms colliding and binding together, later commentators identified an additional dimension to his reductionism: *action-at-a-distance* interactions between the atoms. Sextus Empiricus, a philosopher writing in the late second century, specified that Democritus' atomic theory included a force of attraction between atoms acting across space, drawing them together into composite bodies under a universal principle of like entities attracting one another. The concept of action-at-a-distance force would remain relevant within the physical sciences into the 1800s and cause increasing discomfort for physicists trying to extend their theories of nature to explain how, for example, a force like gravity can reach across the vastness of the solar system and guide the motion of the planets. It would only fall out of vogue with the advent of the field paradigm and, ultimately, Albert Einstein's

---

[4]Textbook author David Griffiths represents a dissenting voice from this sentiment toward honoring Democritus' insights. In his *Introduction to Elementary Particles* (Weinham, Wiley-VHC, 2004), he writes about ancient Greek natural philosophers: "Apart from a few suggestive words their metaphysical speculations have nothing in common with modern science, and although they may be of modest antiquarian interest, their relevance is infinitesimal" (p. 11). This is, perhaps, intended as an uncharitable pun.

[5]Jonathan Barnes, *Early Greek Philosophy*, Penguin Classics (Harmondsworth, Middlesex, England; New York, U.S.A.: Penguin Books, 1987), 247.

redefinition of the structure of space and time in the theories of relativity.[6]

The Roman natural philosopher Lucretius advanced his own reductionist atomic model in the first-century BCE, envisioning all matter constructed of fundamental "bodies" separated by voids. The atomic bodies have two essential properties: their weight, and the ability to obstruct the motion of other bodies. "There is nothing which you can call wholly distinct from body and separate from void, to be discovered as a kind of third nature,"[7] he wrote. Like Democritus' atoms, Lucretius' bodies are indivisible: "These can neither be dissolved by blows when struck from without, nor again be pierced inwardly and decomposed, nor can they be assailed and shaken in any other way."[8] They are uncuttable, solid, and contain no void within them.

The concept of physical reductionism, then, has long been a philosophical starting point for understanding the physics of our world. The atomic theory of the ancient natural philosophers, while amenable to the philosophical avoidance of absurd conclusions, was not without phenomenological support. The permeability of apparently solid matter; the passage of sound through walls; the melting of solid into liquid, which evaporates into gas; the diffusion of dye in a liquid and scent through the air—these are just a few examples of hints from nature to any careful observer, even an ancient one, that the apparent continuity of matter is suspect, and that common sense supports divisibility down to a definite limit.

1.4 Properties of Atoms in Early Physics

The foundation of scientific thinking developed in the Christian West through the Middle Ages and the mechanical philosophy of nature overtook the earlier Aristotelian understanding of teleological causes in the physical world. This developing view of a physical universe

---

[6]We wonder, however, if quantum field theories have smuggled this classical idea back to modern physics, as virtual particle exchanges are postulated to embody interactions.

[7]Titus Lucretius Carus, W. H. D. Rouse, and Martin Ferguson Smith, *De Rerum Natura*, The Loeb Classical Library 181 (Cambridge, Mass: Harvard University Press, 1975), 37.

[8]Lucretius Carus, Rouse, and Smith, 45.

of "nothing but" particles-in-motion lent itself to the Democritian view of those particles in terms of ontological reductionism.

Galileo Galilei took for granted that reducibility has some lower limit, reached in principle by the continued "rubbing together and friction of two hard bodies," which ultimately break apart into *i minimi quanti,* "the tiniest particles" that are "truly indivisible atoms" and distinct in nature from (and perhaps even being the source of) whatever mysterious substance composes light.[9] René Descartes' version of atomism envisioned matter as extension, obscuring the distinction of elementary particles from the surrounding void and from space, itself;[10] but the conceptual agreement between the mechanical philosophy of physics and ultimate parts of nature became increasingly well established. Isaac Newton espoused not only an atomic theory of matter but also a similar corpuscular theory of light, thus envisioning all natural phenomena as reducible to the causal interactions of elementary entities moving through space exerting forces on one another at a distance and through contact.

The late eighteenth-century chemists, including John Dalton,[11] continued to advance a reductionist atomic theory in their field. Although historians of science debate the influences for Dalton's atomic ideas, his reductionism was certainly in the tradition of that held by the ancients and was also clearly guided by experimentation. "All bodies of sensible magnitude, whether liquid or solid, are constituted of a vast number of extremely small particles, or atoms of matter bound together by a force of attraction, which is more or less powerful according to circumstances,"[12] he wrote in 1808. Dalton reasoned that the chemical compounds of nature are constructed from combinations of "simple" elemental atomic species, members of a species being identical to one another, "perfectly alike in weight, figure, etc."[13]

---

[9]Galileo Galilei, "The Assayer," in *The Controversy on the Comets of 1618,* trans. Stillman Drake (London: Oxford University Press, 1960), 313.

[10]Edward Slowik, "Descartes' Physics," ed. Edward N. Zalta, *The Stanford Encyclopedia of Philosophy,* 2014, https://plato.stanford.edu/archives/sum2014/entries/descartes-physics.

[11]John Dalton (1766–1844), an English chemist and schoolteacher, remembered especially for his atomic theory and contributions to early modern chemistry. He also contributed to meteorology.

[12]John Dalton, *A New System of Chemical Philosophy* (Manchester: William Dawson and Sons, 1808), 141.

[13]Dalton, 143.

Dalton thought of atoms as "ultimate particles,"[14] each surrounded by an atmosphere of caloric, the hypothetical substance of heat, which was imagined to exert a repulsive force on neighboring atoms much like the atomic electrons of later models. The simplest and smallest atomic species is hydrogen, and other "simples" are identified as oxygen, nitrogen, "carbone," sulfur, phosphorus, and the metals,[15] varying by weight and diameter. As far as atomic indivisibility is concerned, for Dalton this was an empirical conclusion: "No new creation or destruction of matter is within the reach of chemical agency. We might as well attempt to introduce a new planet into the solar system, or to annihilate one already in existence, as to create or destroy a particle of hydrogen."[16] Even so, "we do not know that any one of the bodies denominated elementary, is absolutely indecomposable; but it ought to be called simple, till it can be analyzed."[17] From this it seems clear that in Dalton's view, as with the ancient philosophers, the bottom level of physical reality would ultimately be identified by *indivisible* bodies that might, nonetheless, come in a variety of forms, distinguishable by intrinsic properties.

Many decades later into the nineteenth century, the great Scottish physicist James Clerk Maxwell[18] penned the *Encyclopedia Britannica's* ninth edition entry on "Atom." "In modern times," he wrote,

> the study of nature has brought to light many properties of bodies which appear to depend on the magnitude and motions of their ultimate constituents, and the question of the existence of atoms has once more become conspicuous among scientific inquiries.[19]

Maxwell emphasized that the discoveries of nineteenth-century physics and chemistry had enriched scientists' understanding of what atoms must be like:

---

[14]Dalton, 209.
[15]Dalton, 222.
[16]Dalton, 212.
[17]Dalton, 222.
[18]James Clerk Maxwell (1831–79) was a Scottish scientist, remembered for his contributions to electrodynamics, thermodynamics, and kinetic theory.
[19]James Clerk Maxwell, "Atom," in *The Scientific Papers of James Clerk Maxwell*, vol. 2 (New York: Dover Publications, 1890), 446.

The small hard body imagined by Lucretius, and adopted by Newton, was invented for the express purpose of accounting for the permanence of the properties of bodies. But it fails to account for the vibrations of a molecule as revealed by the spectroscope,

referring to atomic line spectra. He continued:

The conditions which must be satisfied by an atom are—permanence in magnitude, capability of internal motion or vibration, and a sufficient amount of possible characteristics to account for the difference between atoms of different kinds.[20]

What is evident from Maxwell's encyclopedia entry is that the reductionist theory of indivisible atoms had been refined through experimental means. The building blocks of nature were no longer understood to be necessarily simple: They possess internal properties. Some of these properties have fixed values, which contribute to the absolute distinction between kinds or species of atom, while other properties have fluid or changing values but remain inseparable from the physical structure of the atom.

Ontological reductionism continued to guide early modern physics, receiving encouraging support from experiments. J. J. Thomson's[21] 1897 study of cathode rays and Robert Millikan's[22] 1909 oil drop experiments indicated that electrically charged bits of matter can be broken off and detected as the "electrons" predicted by George Johnstone Stoney in 1894 as the indivisible building block of electric charge. Stoney's reductionist, atomistic speculation had physical support.

Ernest Rutherford's[23] 1911 analysis of scattering experiments identified the atomic nucleus as a small, charged center at the heart of matter filled with empty space, orbited by planet-like electrons. The discovery, a few years earlier, of radioactive splintering of atoms

[20]Maxwell, 470.

[21]Sir Joseph John (J. J.) Thomson (1856–1940), an English physicist known not only as the discoverer of the electron and for the charge-to-mass ratio determination, but many other physics theories and inventions.

[22]Robert Andrews Millikan (1868–1953), an American physicist known also for one of the first determinations of Planck's constant.

[23]Ernest Rutherford (1871–1937), a New Zealander by birth, won the Chemistry Nobel Prize in 1908 for his investigations of radioactive decay. He devised the concept of the *half-life* of radioactive substances. Element 104 of the periodic table is named in honor of him: *rutherfordium*.

may have provided, in Rutherford's words, "a rude shock"[24] to the physical atomic model, but his own research and analysis of matter simply moved the level of elementarity down one level from the chemical atoms to their underlying components, and thus physical reductionism was preserved.

Rutherford's model of the chemical atom differentiated between two species of candidates for the basic building blocks of matter: nuclei and electrons. He would further advance the classical atomic concept by hypothesizing the proton and neutron as building blocks of the nucleus. His student, James Chadwick, provided experimental verification of the neutron in 1932, adding further strength to the theory of ontological reductionism.

The discovery of neutrons might have proved the final step in understanding the elementary structure of matter in our universe: the culmination of the physical reductionist vision. The atomic model of a small positively charged nucleus containing protons and neutrons held together by a strong force of nuclear attraction, surrounded by swarming negatively charged electrons in a kind of quantum mechanical orbit, provides a sufficiently robust model of the composition of matter. With this picture, physicists and chemists can classify the elements of the periodic table and begin to explain the properties of composite structures made up of many atoms held together in solid form. The ordinary matter of our world can be explained with great consistency and usefulness by understanding the structure and behavior of atoms in terms of these three building-block constituents.

1.5   The Descent into the Quark Model

For a moment in the history of science, it had seemed like a level of supreme reductionist simplicity had been finally uncovered. But this victory was short lived as physicists began to observe an ever-growing population of transient structures, or entities, appearing in high-energy experiments. Initially, physicists constructed increasingly sophisticated experimental instruments to passively analyze physical condensation trails in vapor due to the passage of

---

[24]Lord Rutherford, *The Newer Alchemy* (Cambridge: Cambridge University Press, 1937), 3.

mysterious cosmic rays arriving near the surface of earth from outer space. In the presence of a magnetic field, these tracks provided objective visible evidence for the existence of new species of short-lived particles, objectively real and clearly different from electrons, protons, and neutrons.

In the late 1940s, following the Second World War, experimentalists collided known particles with atomic targets and examined the detritus of these high-energy interactions. Much like the consequences of striking a piñata with a stick, a beam of high-energy protons or electrons incident on the atomic nuclei of a stationary target yields a cascade of entities, which can be detected and analyzed in terms of their properties, motion, and energy. As technology improved, the incoming beams of particles could be invested with higher energy, yielding higher-energy outputs from the collision. As physicists analyzed the data from their detectors, they were led to surprising conclusions about the kind of "candy" dropping from their subatomic piñata experiments. Unlike the cloud chamber experiments, there were no physical tracks in vapor indicating the passage of a particle, but a proliferation of short-lived resonances appeared in the data. These resonances were indirect evidence that the energy of the collision experiments resolved into new semi-stable entities that could be distinguished from one another and classified into identifiable types called "baryons" and "mesons" based on their reconstructed properties.

These new entities were too short lived to form together into enduring large-scale structures like chemical atoms, but the consistency of their appearance in the data of collision experiments, with well-defined properties, demanded an explanation. A generation of physicists trained in the reductionist success of the electron–proton–neutron model of the chemical atom naturally applied the same thinking of composite entities to sort out this new problem in terms of some reductionist model of underlying order that could account for the consistency behind this particle zoo.[25]

In a popular science book, theoretical physicist Kenneth Ford commented on one important consequence of this discovery of so many new particles in the 1950s and 1960s:

---

[25]Andrew Pickering, *Constructing Quarks: A Sociological History of Particle Physics* (Chicago: University of Chicago Press, 1984), 12.

Physicists had stopped calling the particles 'elementary' or 'fundamental.' There were just too many of them for that. Yet just as the number of particles seemed to be getting out of hand, physicists were coming up with a simplifying scheme. A manageably small number of particles appeared to be truly fundamental.... Most of the known particles, including the old familiar proton, were composite—that is, built from combinations of the fundamental particles.[26]

A natural model appeared in 1956, proposed by Shoichi Sakata. "It seems to me," he wrote in a letter to the journal *Progress of Theoretical Physics*, "that the present state of the theory of new particles is very similar to that of the atomic nuclei 25 years ago." The discovery of the neutron had allowed an earlier generation to "reduce all the mysterious properties of atomic nuclei to those of the neutron contained in them." Sakata reasoned that a "similar situation is realized at present" and proposed a model to describe the growing number of unstable particles in terms of a small number "of fundamental particles in the true sense:" the already well-understood proton and neutron, an additional building block called the "lambda along with their antiparticle partners."[27]

Sakata's proposal, though given serious consideration by the physics community, was eventually replaced by what was felt to be a simpler and more elegant hypothesis working along the same reductionist assumptions. Murray Gell-Mann and George Zweig independently proposed that the multitudes of baryons and mesons are composite entities constructed from three kinds of "quarks," a building block of matter with the unprecedented property of fractional electric charge. This peculiar property represented a major departure from existing ideas.

Naturally, experiments were immediately conducted to identify free fractional charges moving about in nature, but these turned up empty handed. This awkward absence of direct experimental evidence for quarks has continued down to the present day. It was, initially, a cause for concern regarding the reality of a particle species that should stand out, and even the reductionist potency of the quark hypothesis did not automatically lead to the particle physics

---

[26]Kenneth William Ford, *The Quantum World: Quantum Physics for Everyone* (Cambridge, Mass: Harvard University Press, 2004), 5.

[27]S. Sakata, "On a composite model for the new particles," *Progress of Theoretical Physics*, **16** (n.d.): 686.

community accepting quarks as the actual building blocks of nature. A typical example of this caution is seen in a text from 1982 on the quark model: "Quarks are very elusive objects, and free quarks have as yet not been observed. Actually, we don't know the properties of free quarks, or if quarks exist at all. But...they are useful."[28]

The importance of this problem for the reductionist vision was ably expressed by nuclear physicist Denys Wilkinson in his Wolfson lectures in the spring of 1980:

> Atoms can be got out of the molecules that they compose; electrons, part of the structure of atoms, can be liberated from atoms; nuclei can be dissociated into their constituent neutrons and protons; why, if quarks are inside neutrons and protons cannot they be got out of them? . . . Do not confuse this with the inverse problem: the fact that something comes out of something does not require that it should have been there in the first place; an electron and an antineutrino come out of a neutron leaving behind a proton, as we have noted, but they were not there to begin with; both were manufactured in the act of radioactive beta-decay of the neutron into a proton; barks come out of dogs but dogs are not made of barks. However, if we say that something exists inside something we have to be able to say why we cannot get it out.[29]

Despite the obvious difficulties, any hesitancy appears to have diminished in the succeeding years. On the one hand, this is certainly due, in part, to an accumulation of indirect evidence for the existence of quarks such as the appearance of some internal structure observed within protons subjected to Rutherford-inspired scattering experiments using electrons. On the other hand, the absence of evidence for free quarks roaming the universe was eventually rationalized by a bold hypothesis that these particles must be forever bound to live in groups by an extension of the strong nuclear force that allows them free movement at close proximity but strongly restricts their independence when they try to move too far apart. With these lines of reasoning in place, the familiarity of the quark concept, coupled with the relentless attraction of reductionist

---

[28]D. Flamm and F. Schöberl, *Introduction to the Quark Model of Elementary Particles* (New York: Gordon and Breach, 1982), 16.

[29]Denys Wilkinson, "The Organization of the Universe," in *The Nature of Matter*, ed. J. H. Mulvey, Wolfson College Lectures 1980 (Oxford : New York: Clarendon Press; Oxford University Press, 1981), 25.

solutions, has probably resulted in a drift toward de facto acceptance of these mathematical structures as real entities deserving the status of ontologically real particles.

Today, this appears to be the consensus view of the physics community, and any serious discourse on the ontological status of quarks appears to lie in the past. It is certainly true, at least, that in popular scientific discourse, there is a confident public assurance that we may accept the quark model as a factual account of our most current view of the world. This conviction is adequately exemplified in the simple declaration of the late cosmologist Stephen Hawking: "We now know in fact that neither the proton nor the neutron are elementary but that they are made up of smaller particles."[30] No warning of ambiguity need be given to the non-specialist.

1.6   Contemporary Catalogue of Physical Things

To summarize, our most current understanding of the basic structure of physical reality may be briefly described in reductionist terms as follows.

At the bottom of physical reality are microentities called quarks and leptons, which are real particles of zero size, thought of as occupying only a point in space.

The electron is a specific variety of lepton, and the proton and neutron are understood to be composite structures made of quarks. What differentiate the fundamental building blocks are various inherent properties, including mass, charge, and more abstractly intrinsic spin, color charge, strangeness, and charm. The leptons have the enviable freedom to appear alone in the universe, but quarks—recalling that fractional charges have yet to be observed in experiments—are subjected to a life of permanent confinement, forever bound in pairs or groups, never experiencing the universe as solitary entities.

Composite objects of matter (atoms, molecules, etc.) are composed of two kinds of quarks (called up and down) along with electrons, through the action of forces. These pushes, pulls, and other transformations occur through the intervention of four fundamental

---

[30]Stephen Hawking, *Is the End in Sight for Theoretical Physics? An Inaugural Lecture* (Cambridge; New York: Cambridge University Press, 1980), 2.

interactions: strong, electromagnetic, weak, and gravitation. The first three of these interactions are described in our theories as different expressions of an underlying unity. They are understood to be "mediated" by entities called gauge bosons that are given the names gluon, photon, W, and Z. These four gauge bosons are considered the smallest reducible quantities (quanta) of a universal field associated with each interaction. The gravitational field, on the other hand, the weakest of these interactions, is hypothetically understood to be mediated by a boson called the graviton, whose actual existence is yet to be confirmed by experimental evidence.

1.7   Have We Reached the Bottom?

In providing this standard account of the structure of matter and interactions, many details and complications are, of course, left out, and certainly no attempt can be made within a couple of paragraphs to answer a question like "how do we know this?" We will consider this important question more in the chapters that follow, and in doing so we intend to offer some critical analysis about the correspondence between our current picture of matter and interactions and physical reality, itself.

One very natural, metaphysical question immediately presents itself to any curious person who might consider the possibility, along with Braibent *et al.*, whether "we have perhaps reached the end...in the sense that in the collisions between two particles, it is not possible to reach higher collision energies, that is, higher temperatures. At this level of knowledge, 'Democritus atoms' have been identified."[31] Have we really reached the bottom with quarks and leptons? Indeed, do we even have physical criteria to answer this straightforward yet scientifically and metaphysically important question? The ancient atomic theory, we must recall, defined the elementary building blocks of matter by the key feature of indivisibility, allowing for these fundamental atoms to retain inherent or internal properties. Have particle physicists identified the "uncuttables" of Democritus, or could quarks and leptons yield to further cutting, if the proper tools were applied in the future?

---

[31]Sylvie Braibant, Giorgio Giacomelli, and Maurizio Spurio, *Particles and Fundamental Interactions: An Introduction to Particle Physics*, Undergraduate Lecture Notes in Physics (Dordrecht; New York: Springer, 2012), 8.

An answer to this question requires, first of all, the recognition that apart from the electron (as detected by J. J. Thomson), the fundamental entities of which our theories speak are not accessible to direct observation in the laboratory. Only their signatures are felt as vague traces in a mass of experimental data, like the smile of an invisible Cheshire Cat. Indeed, the situation is more disconcerting than that: The unseen, supposedly physical, quarks and gauge bosons are abstractly defined within our particle physics theories in terms of mathematical superpositions of even more abstract mathematical objects that have little claim, if any, themselves, to real physical existence. We discuss these ontological uncertainties at several points in the chapters that follow.

We return our attention, however, to the orthodox voice of professional physicists for an answer to the reductionist's question: "Are we there yet?" We hope for an answer that makes sense to someone who is not trained in the nuances of the advanced mathematics that define the abstract construction and behavior of quarks, leptons, and gauge bosons to which we just alluded. There are two natural sources of authority: Textbooks and popular science books are supposed to translate the details of the study of reality into the language of everyday life so that everyone else may share the most current insights of professional scientists. Undergraduate textbooks shape the audience of future science professionals who may eventually acquire the technical skills to probe physical reality on their own; popular books of science provide authoritative guidance for educated laypeople who must rely on the expertise of others to guide their knowledge of reality. The audience for popular science publications includes policymakers, science journalists, and specialists in unrelated fields who, nonetheless, find relevance for their own work in understanding the discoveries of advanced physics.

The first thing we should notice is that the reductionist atomic vision of physics is found alive and well in these sources. It is nearly universally portrayed as the reality out of which matter is constructed. Consider the following selection authored for general consumption by well-known physicist Lawrence Krauss:

> For each and every atom in your body, there is a set of quarks, created in those early instants, to which we could trace your existence if we

had the computational means to do so. The atom of oxygen I breathe now, which helps to give the energy to tap the key to type this word, is as connected to one specific set of quarks created in the primordial baseball as I am to my great-grandfather's great-grandfather. Perhaps more so.[32]

Krauss overlooks the fact that the connection would be tenuous, indeed: During the elapsed eons since the primordial baseball expanded to its present form, the quarks of today's universe have suffered the torturous ambiguity of existing in constant mathematical superposition and fluctuation, transformed from one abstract mathematical entity to another. The ontological reality of atoms, nucleons, and their internal quark constituents is, nonetheless, forcefully and authoritatively pronounced by the specialist to his reading audience: there is no doubt that our quarks are real and enduring.

Similar confidence is expressed in the 2013 popular book *Beyond the God Particle* by Nobel laureate and experimental particle physicist Leon Lederman and theorist Christopher Hill: "The elementary constituents of the hadrons are the quarks and gluons. Quarks and gluons are real," they explain, "and their properties are measured, but they can never be set free from the prisons of the hadrons that they comprise."[33]

The reductionist vision of physical reality is an indispensable storyline for this kind of literature, providing a compelling and straightforward framework to explain the discoveries of modern physics. The discovery of quarks as real things provides an opportunity for popularizers to reinforce the success of the modern physics enterprise within the reductionist narrative, and to speculate on the next great discovery awaiting practitioners of that industry. In the 1980s, physicists' reductionist instincts naturally extended to speculation beyond the quark level of matter. The final paragraph in Halzen and Martin's *Quarks and Leptons* (1984) provides a typical example:

---

[32]Lawrence Maxwell Krauss, *Atom: An Odyssey from the Big Bang to Life on Earth... and Beyond* (Boston: Little Brown & Company, 2001), 45.

[33]Leon M. Lederman and Christopher T. Hill, *Beyond the God Particle* (Amherst, New York: Prometheus Books, 2013), 257.

The most important achievement of the last two decades is not just to have established that our world is made of quarks, leptons, and gauge bosons, but to have brought us toward a new frontier where even more exciting questions can be raised. These speculations do inevitably include the possibility that quarks and leptons are themselves composites.[34]

Such anticipation, however, appears to have subsided in later years, inversely related to a growing confidence and satisfaction with the quark model, and a shift in interest toward other matters such as the pursuit of the Higgs boson. Undergraduate textbooks in physics and physical science are typically reserved in approaching these kinds of big questions. A college physics text from 2006 provides a cautious, but ultimately confusing, message that

physicists now believe that leptons are genuinely fundamental particles. The hadron family, by contrast, is a mess.... It has become clear that these particles do not represent the most fundamental level of the structure of matter; instead, there is at least one additional level of structure,[35]

meaning quarks. The implication of "at least" seems to be that quarks might just be composites. Nevertheless, a summary table in the text removes this possibility by affirming quarks as "fundamental building blocks"[36] of everyday substances.

A different text, written in the tradition of teaching physics from a less technical, more conceptual philosophy, is more certain: "Besides the quarks, only the leptons are still elementary that is, not made up of any more fundamental particles as far as we know."[37] The consensus of modern physics is thus portrayed for the emerging student: Quarks are as real as electrons, and they share the ontological bottom.

---

[34]F. Halzen and Alan D. Martin, *Quarks and Leptons: An Introductory Course in Modern Particle Physics* (New York: Wiley, 1984), 354.

[35]Hugh D. Young, Robert M. Geller, and Francis Weston Sears, *Sears & Zemansky's College Physics*, 8th edn. (San Francisco: Pearson/Addison Wesley, 2006), 1034.

[36]Young, Geller, and Sears, 1043.

[37]W. Thomas Griffith, *The Physics of Everyday Phenomena: A Conceptual Introduction to Physics* (Boston: McGraw-Hill, 2004), 448.

1.8   Defining the Bottom Rung of the Ladder

It is important to note, again, that these representative sources assume and affirm physical reductionism in teaching their readers about nature. What we still need, however, is a careful description of how we would recognize a bottom level to the structure of matter, which brings us back to the old idea of uncuttability. A text from 2007 is clear on this line of reasoning, equating the concepts of elementarity with indivisibility, in agreement with the classical ideas: "What is the structure of the physical world on the smallest scale?... To answer (this question) we must turn to particle physics and the search for the elementary, indivisible building blocks of matter."[38]

If elementarity is characterized by indivisibility, then this shifts the burden to identifying the hallmarks of that latter property. How would we know if a particle cannot be subdivided? Physics authors address this through a reasonable metaphysical assumption: Composite entities should have some kind of evident internal structure. Thus, in a world composed of ontological (Democritian) atoms, internal simplicity or homogeneity is a sufficient condition for indivisibility of a particle.[39]

It is not difficult to imagine an extreme limiting case wherein a particle is definitely characterized by internal simplicity and is thus uncuttable: if the particle has zero size. A particle of zero size would certainly be elemental, since one cannot cut something that contains no space in which to set the blade. The ancient Greek natural philosophers would have raised an obvious objection that real things should at least occupy some space. From a modern perspective, however, physics students are now trained from their earliest days to be comfortable with the concept of point-sized real entities, continually acclimatized to the idea as textbook problems ask them to "consider a point particle" as a way to simplify an otherwise difficult or intractable mathematical problem. We thus

[38]Hans C. Ohanian, John T. Markert, and Hans C. Ohanian, *Physics for Engineers and Scientists*, 3rd edn. (New York: W.W. Norton & Co, 2007), 1397.

[39]It is certainly possible to conceive of a fundamentally homogenous substance being divided in parts, but we must remember that the kind of reductionism that is communicated by scientists to the public involves elementary *particles*, not fluids. So long as scientists are looking for particles in the next layer down, they are presumably looking for non-homogenous structures separated by gaps.

find the concepts of elementarity, indivisibility, internal simplicity, and "pointlikeness" intertwined within texts and popular physics writing. A few examples will suffice to show how these concepts are discussed.

A 2005 physics text for scientists and engineers:

> Decades of experiments with electrons and muons, at ever increasing energies, have revealed no evidence that these particles, the leptons, have a substructure. As far as we know, the [leptons and neutrinos] and their antiparticles are 'pointlike.' Pointlike is a good way to describe a fundamental particle; it says that the particle is not 'made' of something more basic, in the way that atoms are made of electrons and nucleons.[40]

Lederman and Hill: "With quarks and gluons a more Democritus-styled explanation of the hadrons took hold, and this is the view that we have of them today. Quarks, like their sisters the leptons, are point-like and structureless matter particles."[41] The authors' interesting invocation of Democritus certainly implies fundamental indivisibility as a necessary property of elementary building blocks, but their redundant equivalence of "point-like" and "structureless" form should be seen as a divergence from actual Democritian atomic theory.

A popular science book, *Rainbows, Snowflakes, and Quarks*, by physicist Hans Christian von Baeyer: "[Quarks] are points and therefore fundamental in a way that protons never were. Their lack of structure lends tentative hope to the proposition that they may escape the fate of being subdivided in the future."[42] This argument is curious in that it prioritizes indivisibility as the key criterion for identifying quarks as more fundamental than protons, yet ignores the fact that protons, themselves remain indivisible by virtue of quark confinement.

In their physics text, Walker and Halliday offer a contrast in that they associate fundamentality with internal homogeneity and do not explicitly invoke infinitesimal size: "In terms of what we know today,

---

[40]Paul M. Fishbane, Stephen Gasiorowicz, and Stephen T. Thornton, *Physics for Scientists and Engineers*, 3rd edn. (Upper Saddle River, N. J.: Pearson Prentice Hall, 2005), 1244.

[41]Lederman and Hill, *Beyond the God Particle*, 257.

[42]Hans Christian Von Baeyer, *Rainbows, Snowflakes, and Quarks: Physics and the World around Us*, 1st edn. (New York: Random House, 1993), 142.

the quarks and the leptons seem to be truly fundamental particles having no internal structure."[43] Later on they write, "electrons and the other leptons seem to be indivisible bodies with no internal structure."[44]

1.9 Are Quarks *Really* at the Bottom?

Atomistic reductionism continues to be the language in which physicists speak to the public, and perhaps to one another, about the ultimate structure of matter. It has not gone completely unchallenged, of course. For example, the Democritian concepts earned criticism from physics luminary Werner Heisenberg, who wrote near the end of his life,

> The atomism of Democritos became an essential part of the materialistic interpretation of the world during the last century: Easily understood and intuitively plausible, it determined the way of thinking of even those physicists who insisted on not dealing with philosophy.[45]

Heisenberg suggested that the very concepts of "divisibility" and "consist of" lose their meaning at the level of the subatomic world and, naively applied by particle physicists, "lead to developments that do not fit the real situation in nature." For Heisenberg, the "real situation" is the following: "The particles of modern physics are representations of symmetry groups,"[46] a statement that sounds very much like the basic level of physical reality is anything but physical, instead the mathematical entities of group theory.

Heisenberg's interpretation of the nature of particles appears to be an expression of what might be called "mathematical realism," a metaphysical stance that identifies mathematical constructs as the ultimate ground of reality. In the next chapter, we will explore the relationship between mathematics and modern physics and propose that in the space between mathematical realism and the more common-sense Democritian or Newtonian physical realism, a

---

[43]Jearl Walker and David Halliday, *Fundamentals of Physics*, 8e, extended edn. (Hoboken, NJ: Wiley, 2008), 1234.

[44]Walker and Halliday, 1413.

[45]Werner Heisenberg, "The nature of elementary particles," *Physics Today*, **29**, 3 (1976): 37.

[46]Heisenberg, 38.

new metaphysical category has arisen in the practice of physicists, and especially particle physicists, which we call quasirealism. To put it briefly, the quasirealist inhabits a universe where physical/atomic reality and mathematical reality become confused or perhaps, more charitably put, intertwined.

As we have attempted to demonstrate through a few examples, however, the non-specialist is almost universally persuaded by the literature available at a popular level that atomistic reductionism is the correct conception of reality at the subatomic level. This common-sense stance leaves only the question of how to identify elementarity upon seeing it, and there appears to be consensus among physicists that indivisibility is the key. It is critical to point out that if this is the only criterion for elementarity, then quark confinement renders protons and neutrons as elementary particles, along with electrons. The fact that quarks are part of the internal structure of the nucleons, but never separable from them, may be the smoking gun that they are not real. We know about quarks only indirectly and, more to the point, mathematically. They can never be observed on their own in the universe. This inherent shyness is all-too-convenient.

The intrinsic mathematical context of quark theory suggests to us that, with the discovery of protons and neutrons, the early modern physics community had already hit bottom, fulfilling the reductionist vision, leaving later generations to simply sort out the details of what kind of internal features and properties are within these building-block uncuttables. These internal features may be given names and assigned properties, but they cannot qualify as the basic building-block particles any more than the indentations on a penny qualify as a more fundamental unit of currency. At the very least, physicists need to learn to talk about the foundations of matter with greater care.

The habitual characterization of elementary particles as point-like, simple, structureless building blocks is a potential, significant stumbling block to accepting this suggestion. A final word about "pointlikeness" is thus in order.[47] Can we conceive

---

[47]Along with the Greeks, we might raise the question as to whether an object of zero size makes physical sense at all or is best understood as an approximation or mathematical convenience. This interesting question would obscure our main point, however, so we will put it aside.

that physical reality might be built of at least some entities that are simultaneously indivisible and still internally complex? Is it obvious that "pointlikeness" is a necessary condition for uncuttability, or has this assumption strayed outside the bounds of well-behaved scientific practice into the venerable realm of philosophy? To answer that, it seems appropriate to turn to a real philosopher for advice.

Philosopher John Heil has made a helpful distinction between the "substantial parts" and "spatial parts" of an object. Spatial parts are really just the regions of an object in space, while substantial parts are found by division into pieces. A "substance," according to Heil, is not made of substantial parts, though it may have spatial parts. According to this language, we may identify the nucleons as being ultimate substances, without concern for their also having spatial extension and containing spatial parts.[48] "Substances," Heil writes, referring essentially to elementary building blocks of nature, "are not hidden beneath, or masked by, their properties. To encounter a substance is to encounter something that is various ways. An electron is a candidate substance. An electron has a definite mass, charge, spin; these are ways the electron is. The electron is not an assemblage of these properties, they do not constitute the electron, nor is the electron an entity separable from them."[49] If the electron, perhaps point-like, can be an elementary particle with distinct properties, then also the proton and neutron might qualify as well, their spatial extension and internal properties forming what they are without the need to identify ontologically more basic, more elementary components within them, namely a new layer of well-defined entities called quarks.

What then should the reductionist make of "quarks"? We will argue in the following chapters that the details of the contemporary quark hypothesis—the mathematics that is supposed to model their real properties—allow for a different understanding of their nature than is normally admitted to the public by the professional physics community. We will argue that the very elegant and powerful mathematical methods and experimental lines of evidence that have supported the "elementary particle" line of reasoning also support a different interpretation: Quarks are mathematically defined structures that effectively describe internal features of real

---

[48]John Heil, *The Universe as We Find It* (Oxford: Clarendon Press, 2012), 38.
[49]Heil, 285.

nucleons but should not be mistaken for real particles, ontologically equivalent to the leptons.

We will, of course, not simply argue about quarks in this way, as if they are worthy of special treatment. Rather, we will extend our constructive critique to other elementary particle candidates in the Standard Model of physics, which we believe are equally susceptible to skeptical caution as to their actual reality. In this endeavor, we will be guided by our awareness that across many disciplines of physics, scientists employ a useful strategy in which complicated real systems are simplified and usefully described by an invented set of familiar, but fictional parameters and attributes: effective theories and models.

The vision of physical reductionism is an enormously powerful one that continues to serve as the meta-narrative of all inquiry in the discipline of modern physics. This is for good reason, because it has yielded a good return on investment across several centuries. However, it might be time to consider the suggestion that our reductionist quest to uncover the basic building blocks of matter actually reached a successful end some time ago, and the continued work since then has become, increasingly, a futile pursuit of mathematically abstract quasi-entities. To continue, stubbornly, to insist that these increasingly complicated mathematical entities are real things is, we believe, an interpretative mistake and is actually taking us away from the original trajectory of our quest. We will describe this mistake in the next chapter as we explain the key concept of quasirealism.

Chapter 2

Quasirealism

The reasonable man adapts himself to the world: the unreasonable one persists in trying to adapt the world to himself. Therefore all progress depends on the unreasonable man.

—George Bernard Shaw
The Revolutionist's Handbook (1903)

2.1   Common Sense

Common sense, according to the philosopher H. H. Price, is "a body of very general principles commonly accepted by ordinary non-philosophical men in the ordinary affairs of life."[1] One of the most important tasks undertaken by philosophers is to critically examine the assumptions of common sense and determine if these are truly consistent. This has been true of philosophers since at least back to the time of the famous Athenian, Socrates. Every student knows of Socrates' famous strategy of challenging common sense by asking a series of carefully composed questions, which ultimately reveal the flaws in otherwise obvious opinions. However, where philosophers like Socrates *challenged* the everyday thinking of their peers, *natural* philosophers (early scientists) *relied upon* common sense to understand the physical world.

In an idealized sense, the scientific method begins with intuitive explanations of natural phenomena, which are then explored

[1]H. H. Price, "The appeal to common sense," *Journal of Philosophical Studies*, **5**, 17 (1930): 24.

From Atoms to Higgs Boson: Voyages in Quasi-Spacetime
Chary Rangacharyulu and Christopher Polachic
Copyright © 2019 Jenny Stanford Publishing Pte. Ltd.
ISBN 978-981-4800-24-4 (Hardcover), 978-0-429-02765-9 (eBook)
www.jennystanford.com

through a series of questions and hypotheses, tested by careful experimentation. Failed hypotheses are discarded, while successful ones are further developed and more intensely scrutinized, probing deeper into how Nature works. Prior to the nineteenth century, natural philosophers appealed to common convictions and observations in their quest to understand the structure and rules of Nature, and over the centuries this approach met with remarkable success even before the advent of systematic experiments in laboratory conditions. "This," Werner Heisenberg wrote, "simply shows how far one can get by combining the ordinary experience of Nature that we have without doing experiments with the untiring effort to get some logical order into this experience to understand it from general principles."[2]

As the scientific method and humanity's knowledge about Nature developed over time, certain basic concepts were held to be self-evident. Thus, while Socrates reasoned with and dissected the common-sense ethical views of his audience, his contemporary Democritus appealed to their common sense in his theorizing about the fundamental physical structure of the world. Contrary conclusions were ultimately dismissed by the natural philosophers as "utterly silly" and "absurd" because certain concepts about physical reality are clear to every thoughtful person who can follow the details.

In more recent times, the virtue of common sense has become increasingly devalued in certain scientific disciplines. Physicists, in particular, have become acclimatized to the assumption that everyday notions are grossly insufficient to describe the working of physical reality in a fundamental sense. This is, of course, the result of the magnificent success of Einstein's relativity and quantum theory, the two pillars of twentieth-century physics that established reasoning contrary to common sense as *de rigueur*. Physicists are now trained to accept violations of common sense in these theories as a point of celebration. A college course in modern physics is taught with the proviso that the truths of our world may not make sense to the student. We are warned that our classical, macroscopic instincts break down at the quantum level: Nature is stranger than fiction on very small or fast scales. Physics popularizers seem to delight in the license of describing relativity and quantum theory in terms such as "bizarre," "spooky," "counterintuitive," "eerie,"[3] and "baffling."

---

[2]Werner Heisenberg, *Physics and Philosophy: The Revolution in Modern Science*, Great Minds Series (Amherst, N. Y.: Prometheus Books, 1999), 75.
[3]Kenneth William Ford, *The Quantum World: Quantum Physics for Everyone* (Cambridge, Mass: Harvard University Press, 2004), 247.

2.2   Mathematics at the Centre

This move to embrace models of physical reality that defy our common sense has come hand-in-hand with the physicist's conviction that mathematics is a reliable guide for understanding the structures and interactions of our world. The rationale for this faith is explained by physicist Eugene Wigner: "We have seen that there are regularities in the events in the world around us which can be formulated in terms of mathematical concepts with an uncanny accuracy."[4] Wigner went so far as to assert that "the laws of nature must already be formulated in mathematical language."[5] Particle physics theorist Paul Dirac exemplified a more extreme form of this mathematical fundamentalism in proposing "a rather general principle in the development of theoretical physics," that "one should allow oneself to be led in the direction which the mathematics suggest...and see what its consequences are, even though one gets led to a domain which is completely foreign to what one started with."[6] The means justify the end, however strange it might be.

The roots of this perspective go back at least as far[7] as Galileo Galilei, who laid the foundations of a mathematical physics in the seventeenth century. Galileo considered mathematics to be "the language" of the universe. Nature "cannot be understood unless one first learns to comprehend the language and interpret the characters in which it is written." His view of this language, however, was limited to common-sense mathematical forms such as "triangles, circles, and other geometrical figures," which are required in our conceptual toolkit in order to avoid "wandering about in a dark labyrinth" of speculation and opinion rather than facts.[8] For Galileo, mathematics was a tool that ensured consistency while one explored what is

---

[4]Wigner, Eugene P., "The unreasonable effectiveness of mathematics in the natural sciences," *Communications on Pure and Applied Mathematics,* **13**, 1 (1960): 11.
[5]Wigner, Eugene P., 6.
[6]Paul Dirac, "The origin of quantum field theory," in *The Birth of Particle Physics*, eds. Laurie M Brown, Lillian Hoddeson, and International Symposium on the History of Particle Physics (Cambridge [Cambridgeshire]; New York: Cambridge University Press, 1986), 41.
[7]There are certainly ancient tendrils of this perspective among the classical Greek thinkers, such as Pythagoras and Plato.
[8]Galileo Galilei, "The Assayer," in *The Controversy on the Comets of 1618*, trans. Stillman Drake (London: Oxford University Press, 1960), 184.

known and not a guide *into* the unknown. The connection between the structure and rules of Nature and mathematical concepts took root and grew vigorously. Succeeding generations of physicists have applied increasingly complex mathematics to the task of physical modeling, without apparent limitations on the success of such a synthesis.

Only a few decades after Galileo, Newton's development of calculus was essential to his work on motion, forces, optics, and gravity. In the nineteenth century, Maxwell's differential equations of the electromagnetic field provided a mathematical justification for both the field theory of Nature's fundamental forces of interaction and for Einstein's postulate, half a century later, that the speed of light is constant in all inertial reference frames.

Throughout the nineteenth century, physicists learned to deal with otherwise intractable complex systems of many parts and many degrees of freedom using statistical methods, providing insight into abstract macroscopic properties where the real microscopic details remained out of reach. For example, when we are concerned with the evolution of the temperature, pressure, and volume of a gas in a container, we refer only to the average and maximum values of momenta and kinetic energies of an ensemble of entities, along with a general distribution formula. There is no need for exact values of momentum or position with respect to each atom or molecule in the container. Statistical methods emboldened physicists to see ignorance of the physical details not as a handicap, but an opportunity to reinterpret the physical parameters of interest.

In formulating general relativity, Einstein moved away from the Newtonian conviction that masses and momenta of entities are causal physical properties in the dynamics of Nature and instead described the essential physics of systems in terms of energies and fields in a curved spatial geometry. Following the lead of others before him, he used *space* and *geometry* as synonyms, embracing the deepening conviction that an effective mathematical treatment of a natural system has priority over a common-sense interpretation of the model's details.

By 1930, the ontological identity of matter and the rules governing the universe on its smallest scale would be fully redeveloped along mathematical lines in the quantum theory. With that as a starting point, and the statistical methods of nineteenth-century thermodynamics as inspiration, solid-state, atomic, and subatomic physicists discovered that complex systems could be mathematically tamed by carefully redefining physical parameters in their equations, resulting in tidy solutions. In these schemes, the true physical parameters of the complex system being modeled are replaced with averages, *mean fields*, and abstract mathematical substitutions, which provide tremendous simplification and economy of expression on the theorist's blackboard. Out of these solutions arise mathematical entities that have properties reminiscent of those of the real particles and interactions that lie beneath. These new entities are appropriately referred to as *quasiparticles*.

2.3 Quasiparticles

Solid-state physicists use the quasiparticle paradigm to great effect and are generally quite careful in remembering the limits of this technique. For example, the mathematical model of a crystalline solid replaces free (quantum mechanical) electrons with new quasielectrons whose wave function is redefined in terms of the overall, spatially distributed influence of the solid material's periodic atomic structure. These new quasiparticles are described by quantum mechanical *Bloch* and *Wannier functions*, which effectively replace all the complicated interactions of the crystal environment with the appearance of free particles traveling through an artificial space with somewhat different physical rules. In this new, fictional universe, quasiparticles may have fractional electric charge, or a mass several hundred times that of real electrons, or even acquire negative mass under certain conditions.

Solid-state quasiparticle descriptions are enormously useful, as in the case of Cooper pairs in BCS theory, which provides an effective mathematical description of the very real phenomenon of conventional superconductivity, where macroscopic electric current propagates through certain materials under special conditions with zero electrical resistance. Solid-state physicists (usually) employ quasiparticle descriptions with the explicit understanding that these are fictitious entities, and the real mechanics of the complex phenomena are epistemologically buried in the complexity of the real system.

This kind of philosophical clarity is not often seen, however, in the world of particle physics where the mathematical entities of the most recent theories are normally identified as real entities of the physical world. Particle physics paradigms have become so entangled in mathematical nuance that it is impossible for the non-specialist to picture the claims being made about our world without a generous amount of cartoon illustration on the part of experts, always to the effect that the cartoons are basically a true representation of reality.[9]

Historically, nuclear and particle physicists borrowed on the successes of quasiparticle solid-state physics and constructed mathematical spaces in which their entities propagate and interact, leaving apparent signatures in the laboratory. Experiments are now designed, analyzed, and interpreted by these practitioners to support their theoretical quasiphysics, and rarely (if ever) is the public reminded that these outcomes deal with a quasi-Nature filled with quasientities. As we discuss in later chapters, the elementary mathematical entities in this approach are conveniences with transitory attributes that are very often far removed through multiple stages of calculation and interpretation from the ontologically real, physical matter or energy.

It is important to emphasize that particle physicists are the specialists we rely upon to provide insight into the structure of the most fundamental scale of our physical universe. Our millennia-long pursuit of the reductionist vision has now become wedded to increasingly complex and arcane mathematics. Mathematical modeling in particle physics is now systematically confused with submicroscopic reality. Nuclear physicist Timothy Smith admits as much in a popular science book on quarks: "The trend in the study of matter is not just a migration to smaller and smaller scales, it is also a migration to more abstract evidence, evidence that can only be understood through a growing reliance upon theoretical interpretation."[10] Non-specialist laypeople and undergraduate students in physics are unable to distinguish quasiphysics from real physics, and the gatekeepers of the appropriate body of knowledge, who should act as guides in these matters, have, seemingly, embraced

---

[9]As an example, particle physicists make generous use of "Feynman diagrams" that conceal immensely complicated mathematical integrals in the misleading guise of trajectories drawn in a two-dimensional spacetime plane.

[10]Timothy Paul Smith, *Hidden Worlds: Hunting for Quarks in Ordinary Matter* (Princeton: Princeton University Press, 2003), 52.

the illusion. It may be safe to say that many are no longer aware of the distinction.

2.4   Quasirealism Defined

In this book, for the purpose of clarity, we introduce[11] a new kind of philosophical position to describe this attitude of implicit trust in partnering mathematics with the historical reductionist quest of physics: *quasirealism*. In short, quasirealism is a philosophical perspective on science that identifies otherwise clearly unphysical mathematical conveniences as having real existence in the physical world.

Note that in quasirealism there are aspects of a theory that are understood to be referring to real physical things, but the interpreter *should know better*. The astute physicist should keep her eyes open for mathematical manipulations that, although useful, render the final form of the theory and its direct referents as no longer containing direct relevance to physical reality. The theory or model will have evolved out of a series of abstractions, generalizations, transformations, redefinitions, or statistical procedures that render it only indirectly connected back to the physical system of interest. Thus, a quasirealist view of a physics theory is different from the realist belief that the formal elements of the equations correspond to real, physical things in Nature. That should often be the case in physics, especially when equations describe simple systems. But quasirealism involves the *misattribution* of reality to formal parts of a theory due to the practitioner losing sight of what they were doing.

Quasirealist interpretations tend to grow with time as physical theories are passed on from teacher to student. The best of teachers will inform their students of all the nuances out of which the body of knowledge has developed, highlighting those theoretical aspects

---

[11]So far as we can tell, the term *quasirealism* has not been used before in the context of philosophy of science in quite this way. The term *quasi-realism* has been used by philosopher Simon Blackburn in the context of moral philosophy, and our use has—it seems to us, at least—only incidental similarity to his. See, for example, Joyce, Richard, "Moral anti-realism," *The Stanford Encyclopedia of Philosophy* (Winter 2016 Edition), Edward N. Zalta (ed.), URL = https://plato.stanford.edu/archives/win2016/entries/moral-anti-realism/ (accessed March 2018). Any study of deeper consilience between our *quasirealism* and Blackburn's *quasi-realism* would certainly require the work of a proper philosopher.

that do not directly correspond to physical observables. The student struggles, initially, to simply master the mathematical details of the theory and, in the relief or triumph of this accomplishment, has forgotten the foundational context. The student, graduating into a professional teacher or practitioner, tends now to feel comfortable with the abstractness of the theory and uses it without adequate critical reflection on what it really means.

This is especially easy if the physical system of interest occupies a scale of matter that is beyond direct perception and verification, which is most clearly the case in the world of particle physics. It is for this reason, we believe, that although quasirealist errors of interpretation may be found in all branches of the physical sciences, they are especially acute in that sub-discipline in which the limits of reductionism are aggressively—and competitively—pursued.

Quasirealist interpretation is exacerbated by the pressure of public interest on these scientists who receive large amounts of public funds to pursue research into questions of truth that are the subject of universal curiosity. The inaccessibility of this scale of physical reality, for the common person, elevates particle physicists to the privileged role of interpreter of obscure (mathematical) holy texts that assuredly contain special insights into the true nature of our world. There are uncanny similarities between modern physics and ancient Gnosticism that generate a culture in which quasirealism thrives, much to the detriment of the reductionist quest.

It is probably fair to say that most physicists do not worry much about the philosophical details of their study of Nature. However, most will necessarily operate under the assumption of what philosophers have called scientific realism. This is the perspective that the fields, particles, and interaction forces that are postulated in our theories are actually symbolic of something that is ontologically existent. It, thus, entails a commitment to the philosophical conviction of physical realism,[12] the view that there is a physical reality in Nature that exists regardless of what we might have to say about it.

---

[12]This has otherwise been called *materialism* in contrast to *idealism*. However, the term "materialism" is also used for the view that *only* material entities exist in reality, as opposed to other kinds of ontological categories such as *spiritual* entities. We have not attempted to be overly thorough with these technical details. As non-philosophers, we admit that we can neither understand nor capture every nuance that is appropriate to a rigorous discussion of philosophy of science and thus, as humble scientists, we beg for mercy before the more sophisticated judgment of our philosopher colleagues who may stumble upon these pages.

Realist physicists are not naïve: We acknowledge that our current theoretical entities are not perfect descriptions of the real ones. However, one theory is better than another just insofar as it gets us closer to a proper description of what is really there in Nature. It is the conviction that a real thing exists, awaiting accurate description that makes us realists. It is the job of the scientist to systematically uncover this reality, over time, through the careful application of theory and experiment, confident that there is only one right answer in the end. Physical realism is a conviction ably expressed by Robert Millikan:

> To the experimental physicist, at least, this world is at bottom more than a world of equations or even ideas. Some external physical things are happening and we cannot rest indefinitely content with two types of physical interpretation of the same phenomena that seem to be mutually exclusive.[13]

A very different perspective from physical realism is that of phenomenalism, a view that grounds certainty about reality at the level of our sense perceptions. Idealism is a more extreme position locating reality at the level of our mental constructs. In both of these metaphysical perspectives, the world of physics is nothing but our personal world of sense or thought. This is the reverse of a robust physical reductionism that would argue for our biological and mental sense perceptions as (at least to a partial extent) outcomes of particles in motion!

Non-physicalist philosophies such as phenomenalism and idealism make for entertaining discussion over dinner but seem like more trouble than they are worth when life goes on in the laboratory observing particle tracks and discerning resonances in a data set: The consistency of results from one experiment to another surely indicates a level of physical reality independent of our perceptions. Even so, some famous physicists have apparently espoused such views. Hermann Weyl[14] was among them. Inspired by the successes of Einstein's relativity, Weyl argued that physical concepts of space, time, and matter, commonly understood as the "forms of existence" and "substance" and "the real world," are inadequate concepts for understanding reality. "And now, in our time," he wrote in the years

---

[13]Robert Andrew Millikan, *Electrons (+ and –), Protons, Photons, Neutrons, and Cosmic Rays* (Chicago: University of Chicago Press, 1935), 267.

[14]Hermann Weyl (1885–1955) was a German mathematician and mathematical physicist.

following the introduction of relativity and the birth of quantum theory, "there has been unloosed a cataclysm which has swept away space, time, and matter hitherto regarded as the firmest pillars of natural science, but only to make place for a view of things of wider scope, and entailing a deeper vision."[15]

Space, time, and matter, to Weyl, are subjective concepts, "intentional objects" that attain their full reality only as they are mentally internalized by a conscious observer of sense perceptions, interpreted through mathematical physics.[16] Indeed, for Weyl the mathematician, not only are colors, as sense perceptions, not real in themselves (being subjective), they cannot even be described more accurately as wavelengths of electromagnetic waves: "colours are thus 'really' not even aether-vibrations, but merely a series of values of mathematical functions in which occur four independent parameters corresponding to the three dimensions of space, and the one of time."[17]

Weyl's ideas about reality are several steps removed from common sense. The real properties and entities of the universe, and even the more abstract fields that have already fallen out of Maxwell's mathematics of electricity and magnetism, are only real in the sense that they are mathematically described. In this view, any experiment that analyzes extension, duration, or properties of material substance (such as mass) is not "really" informing us about the world in a fundamental way.

We do not criticize Weyl's idealism: It is, along with phenomenalism, a nuanced position that seeks to make sense of reality. Not many practicing scientists would explicitly endorse this view, but a casual flirtation with Weylian idealism opens the door for all kinds of confusion in modern physics. Many physicists today practice compartmentalization in their thinking between physical realism and mathematical idealism: They want physical reality to exist, out there (otherwise, why study physics?), but also enjoy the relative ease of systematically and uncritically interpreting mathematical quasientities as providing direct reference to those real things. The assumption of physical reality is basic in a physicist's thinking, but Weylian idealism is added on for flavor, and the mixture of these two ingredients, in this order, results in a quasirealist interpretation of theoretical statements.

---

[15]Hermann Weyl, *Space–Time–Matter*, trans. Henry L. Brose, 4th edn. (London: Methuen & Co. Ltd., 1920), 1.
[16]Weyl, 2.
[17]Weyl, 4.

2.5 Against Quasirealism

In our view, the antirealist metaphysics of idealism has had a contaminating effect on the way physicists of the present day view their theories, and quasirealism has become the received wisdom uncritically endowed by physics teachers onto their students. The influence of thinkers such Weyl and Dirac has not, however, totally saturated all of physical thought in the last hundred years. Contrary views are certainly to be found among some of the most influential voices. However, to the ears of physicists reared in the halls of modern undergraduate physics programs, criticism of quasirealist assumptions will sound quaint, even heretical.

One natural criticism that might be levelled against our concerns is that we are obstinately yearning for a revival of nineteenth century, classical physics common sense—the kind of stubborn hostility that Albert Einstein is accused of holding against the new quantum theory, expressed in his reluctance to accept that deterministic physics was demonstrably dead. Let it be noted that refusing to interpret abstract mathematics as describing something ontologically real or physical about our universe is not the same as falling back on classical, Newtonian physics as the purest assessment of physical reality. The controversy over the basic nature of light, between Newton's corpuscular theory and Hugyen's wave paradigm, demonstrates that classical theories were fraught with their own ambiguities. We know that classical physics, though in basic agreement with many common-sense perceptions of our world, does not align with modern experimental evidence, and the theories of relativity along with quantum theory provide immensely helpful improvements. Our concern is that we must interpret the statements these theories make about physical reality with care, keeping in mind that one of the goals of physics is to provide a true account of how the world is structured—insofar as that is possible. The inverse of that responsibility is to boldly identify those aspects of our theory that are simply useful fictions.

There are some rather illustrious examples of respected physicists who have resisted quasirealism in their work, without pining for a classical revival. With respect to interpretations of quantum theory, J. S. Bell's personal preference for the deterministic de Broglie–Bohm pilot-wave paradigm, which retains some classical notions of particles, was not a stubborn insistence on strict Newtonian ideas. Rather, Bell appealed to clarity and "craftsmanship." He recognized that,

> in theoretical physics sometimes the inventor knows from the beginning that the work is fiction, for example when it deals with a simplified world in which space has only one or two dimensions instead of three. More often it is not known till later, when the hypothesis has proved wrong, that fiction is involved. When being serious, when not exploring deliberately simplified models, the theoretical physicist differs from the novelist in thinking that maybe the story might be true.[18]

The mathematical details of a pilot-wave quantum theory, in Bell's view, are compatible with common sense in many important ways, and he felt that this line of interpretation only lost out to quasirealist views due to historical ignorance.[19]

Louis de Broglie[20] serves an example of a physicist who, reportedly, rejected quasirealism despite advancing much of the language and conceptual basis on which it is grounded today in his confounding treatment of particles as waves during the opening years of quantum theory. His biographer and colleague, Georges Lochak, wrote of him,

> Louis de Broglie was an intuitive thinker, concrete and realistic, attached to simple physical pictures in three-dimensional space. He refrained from attributing any ontological value to abstract mathematical representations, especially in multidimensional spaces, such as Hilbert space or the configuration space of dynamics. He viewed them only as convenient tools.[21]

---

[18]John S. Bell, *Speakable and Unspeakable in Quantum Mechanics*, 2nd edn., reprint, Collected Papers on Quantum Philosophy (Cambridge: Cambridge University Press, 2008), 194–195.

[19]Or perhaps ignorance grounded by conspiracy: Bell wrote, "the pilot wave interpretation was rather deeply consigned to oblivion by the founding fathers [of quantum theory], and by the writers of text-books" (Bell, 193).

[20]Louis Victor Pierre Raymond de Broglie (1892–1987), a French aristrocrat and history-student-turned-physicist. De Broglie's controversial doctoral thesis introduced into physics the idea that wave–particle duality applies to matter as well as to light. Incidentally, his last name is properly pronounced "de-broy."

[21]Georges Lochak, "Louis de Broglie's conception of physics," *Foundations of Physics*, **23**, 1 (1993): 123.

De Broglie also rejected the real equivalence of space and time in Einstein's relativity, while firmly embracing the usefulness of the theory,[22] which was the inspiration for his own doctoral thesis on electron waves.

Nuclear physicist Denys Wilkinson exemplified the kind of careful thinking that is often ignored in a physicist's quasirealist education, in his description of the van der Waal's force: an effective interaction between atoms caused by interpenetrating electron clouds. The influence of the underlying electrostatic interaction (which obeys the inverse square Coulomb's law) appears to be replaced by a new force that can be mathematically modeled by an inverse seventh-power law, due to the complicated electrostatic interactions among electrons and nuclei. This new force has an interesting mathematical form that provides a convenient example for analysis in undergraduate textbooks, where it often seems to be treated as a basic interaction between atoms. "However," Wilkinson wrote,

> it is in no fundamental sense a force in its own right but only a representation of a particular aspect of the inverse square law of electrostatic force operating within the laws of quantum mechanics between extended and structured electrostatic systems that each possess over-all electrical neutrality.[23]

In the final year of his life, the solid-state physicist Léon Brillouin (whose eponymous Brillouin zones daily remind every graduate student in condensed matter physics that abstract mathematical concepts serve very useful purposes) offered unvarnished criticism of quasirealist interpretations of the general theory of relativity. "General relativity," he wrote, "is a splendid piece of mathematics built on quicksand and leading to more and more mathematics about cosmology (a typical science-fiction process)."[24] Brillouin's disapproval was not about general relativity in terms of internal consistency, but rather the assumption that it provides a clear statement about the nature of Nature, and this father of solid-state

---

[22]Ibid.

[23]Denys Wilkinson, "The organization of the universe," in *The Nature of Matter*, ed. J. H. Mulvey, Wolfson College Lectures 1980 (Oxford : New York: Clarendon Press; Oxford University Press, 1981), 27.

[24]Leon Brillouin, *Relativity Reexamined* (New York: Academic Press, 1970), 10.

physics hoped that the conceptual problems of interpretation would motivate a better theory in the future.

It may be that Niels Bohr's pragmatism, as described briefly in Chapter 7, is the best response available to a physicist confronted by the spectacular success of quantum theory and relativity, but who is also unswerving in the common-sense instinct that the use of imaginary numbers signals the end of a one-to-one relationship between theory and reality. There is no shame in a physicist ultimately admitting that we do not yet truly grasp the underlying structure of physical reality, and that—perhaps—there are some secrets we cannot ever capture in a theory except as shadows of the real thing.

The history of particle physics in the last hundred years has been especially saturated with quasirealist interpretations. Philosopher of science Andrew Pickering insisted, however, that alternative views, similar in spirit to those just illustrated, are certainly credible. In his 1984 sociological study of quark theory's development, Pickering explained that the details—the history—of how the high-energy physics (HEP) community has developed its reductionist model of matter and interactions demonstrate that the orthodox Standard Model is "an understandable version of reality," but not an exclusively true one. Other models competed for attention and would have worked just as well. "Given their training in sophisticated mathematical techniques," Pickering explained, "the preponderance of mathematics in particle physicists' accounts of reality is no more hard to explain than the fondness of ethnic groups for their native language." However, he adds,

> there is no obligation upon anyone framing a view of the world to take account of what twentieth-century science has to say. The particle physicists of the late 1970s were themselves quite happy to abandon most of the phenomenal world and much of the explanatory framework which they had constructed in the previous decade. There is no reason for outsiders to show the present HEP world-view any more respect.[25]

We would re-express what Pickering is saying more cautiously in terms of quasirealism: We can respect the work of our particle physics colleagues as exceptionally useful, even providing some measure

[25]Andrew Pickering, *Constructing Quarks: A Sociological History of Particle Physics* (Chicago: University of Chicago Press, 1984), 413.

of correspondence to the physical world that should be taken into consideration in our ultimate quest for reductionist ontology of matter. But there are good reasons for outsiders, and for students entering a career in particle physics, to question whether the current explanatory framework of the Standard Model—or, indeed, of any variant or extension of that model, which are continually being contemplated—provides as much insight into reality as is typically claimed. The mathematical formulations of the theory have taken us far beyond anything actually observed in the physical laboratory, and the entities that are proclaimed to be fundamental components of reality are, instead, abstract mathematical structures dwelling in multidimensional complex mathematical spaces that have no physical parallel or tangible relevance outside of the effective theories of nuclear and particle physics.

2.6   Quasirealism and the Theory of Everything

In 1936, Albert Einstein was still able to propose that "the whole of science is nothing more than a refinement of everyday thinking."[26] The extraordinary achievements of twentieth- and twenty-first-century particle physics, however, have been driven by mathematical theories divorced from anything familiar to an educated layperson, such as Galileo's triangles and circles. This remarkable progress has brought fundamental science to a place that challenges Einstein's conviction. Ironically, the seeds of quasirealism were planted in large part through the assertions he made about space and time in his special theory of relativity.

Nevertheless, the language used to promote research activities in fundamental physics, and to popularize its discoveries at the level of "everyday thinking," and even in the sphere of undergraduate education, gives the impression that common-sense concepts about Nature are ontologically valid. Physicists communicate their work to the public using classical imagery such as "particle," "force," and "mass." A common conviction about the structure of physical reality lies at the heart of both "everyday thinking" about Nature and the industry of modern physics: It is physical reductionism.

---

[26]Albert Einstein, *Out of My Later Years: The Scientist, Philosopher, and Man Portrayed through His Own Words* (New York: Wings Books, 1993), 59.

Reductionism has two flavors in modern physics, which are certainly related in physicists' thinking, but not quite the same. On the one hand, there is a materialist reductionism, involving a quest to find the ultimate building blocks and interactions of physical reality, as outlined in the previous chapter. This is an ancient idea and one grounded in Democritian common sense. The other kind of reductionism, however, is a mathematical one, grounded in the conviction that all of physics will one day be reduced to a single "theory of everything," perhaps a set of equations or a unifying single equation that is all-encompassing and explanatory.

Albert Einstein characterized this theoretical, mathematical reductionist vision in eloquent and visionary terms, believing in its universal role as a motivation for scientists:

> From the very beginning there has always been present the attempt to find a unifying theoretical basis for all these single sciences, consisting of a minimum of concepts and fundamental relationships, from which all the concepts and relationships of the single disciplines might be derived by logical process. This is what we mean by the search for a foundation of the whole of physics. The confident belief that this ultimate goal may be reached is the chief source of the passionate devotion which has always animated the researcher.[27]

Paul Dirac agreed, stating triumphantly as early as 1929 that the end was in sight for chemistry (certainly) and physics (nearly so):

> The underlying physical laws necessary for the mathematical theory of a large part of physics and the whole of chemistry are thus completely known, and the difficulty is only that the exact application of these laws leads to equations much too complicated to be soluble.[28]

Stephen Hawking reinvigorated the theme in 1980, more hesitantly, in his inaugural lecture as Lucasian professor of mathematics at Cambridge:

> In this lecture I want to discuss the possibility that the goal of theoretical physics might be achieved in the not too distant future, say, by the end of the century. By this I mean that we might have a complete,

---

[27] Einstein, 99.

[28] P. A. M. Dirac, "Quantum mechanics of many-electron systems," *Proceedings of the Royal Society of London A*, **123** (1929), 714–733.

consistent and unified theory of the physical interactions which would describe all possible observations.[29]

The dream of a final, unifying, almost supernatural mathematical theory energizes quasirealism, making it immune to criticism. The prophecies foreseeing the end of science in a single equation support the unexamined belief that the mathematical quasientities of current theories are to be taken literally. And if, on the road to that final theory, some future replacement of the Standard Model dispenses with the mathematical structures interpreted as physical quarks or gauge bosons, one wonders how our scientific descendants will understand the indiscriminate confidence universally expressed today in their ontological status.

2.7   Concluding Thoughts

In practice, or at least in public, the assumption of physical realism is robust among physicists, despite the tendency toward quasirealism. It defies all kinds of red flags thrown at it by the mathematical obscurities and paradoxes that have grown up in the quantum theories of mechanics and fields, and within the two theories of relativity. The insights of these theories are often termed "bizarre" and "spooky" by popularizers.

On the one hand, physicists are propelled into deeper and deeper investigation of the structure of Nature by the confidence that, in some common-sense way, they are investigating what is really there. The physicists' attachment to physical realism is evident in straightforward press releases, particle data tables, and textbook figures that systematize the structure of subatomic reality in concrete definitions of particles and interactions. Quarks and gluons, gravity waves, neutrinos, positrons, and the Higgs boson are somewhere down there, objectively and reliably structuring our world.

On the other hand, there should be no illusion that these are "theory-laden" entities, as will be explained in detail in the chapters that follow. It is an interesting tension that the modern physical scientist must live within and, more often than not, simply ignore. But that is unfair to the public, to policymakers, to journalists,

---

[29]Stephen Hawking, *Is the End in Sight for Theoretical Physics? An Inaugural Lecture* (Cambridge; New York: Cambridge University Press, 1980).

to fellow scientists in other disciplines, and to undergraduate students of physics—all of whom rely on professional physicists to be clear about the limits of our current knowledge. It is even the case that many who work as particle physicists must simply trust the quasirealist pronouncements of their colleagues, practically unable to critically evaluate work within their own discipline. To double-check and objectively evaluate the detailed calculations and experimental assumptions of their colleagues, as we expect in good scientific practice, would require impractical investment of time, computational resources, and experimental expense: resources that, at best, may only be accessible to a select few.

Chapter 3

Space, Time, and Relativity

Science is a beautiful gift to humanity, we should not distort it.

—A. P. J. Abdul Kalam
TIME interview (1998)

Understanding the nature of Space and Time[1] is basic to the success of physics, as any discussion of motion, interaction, cause, and effect and the details of experimental design require that physicists share some objective understanding of how to quantify the position and duration of events in our universe. The subject is a difficult one to address with conceptual clarity, partly because we are immersed in this context like fish in water, and experiential objectivity is elusive. Nonetheless, clarity on this point lies at the heart of any realist view of physics because whatever we say about the nature of Space and Time should remain consistent whether we are talking about gravitational, electromagnetic, atomic, or subatomic processes. It is also a question where quasirealist perspectives play a profound role in the life of modern physics philosophy and interpretation.

To see the role that quasirealism has come to play in understanding the ontological characteristics of Space and Time, it is important that we examine the evolution of these concepts over the history of natural philosophy and the advent of modern science.

---

[1]To avoid confusion, we capitalize Space and Time in this chapter to refer to hypothetical concrete entities that may form a kind of basic, independent background to the physical universe. In this sense, the lowercase time over which an event occurs or the space spanned by a physical entity are specified within the background of our universe's Space and Time.

From Atoms to Higgs Boson: Voyages in Quasi-Spacetime
Chary Rangacharyulu and Christopher Polachic
Copyright © 2019 Jenny Stanford Publishing Pte. Ltd.
ISBN 978-981-4800-24-4 (Hardcover), 978-0-429-02765-9 (eBook)
www.jennystanford.com

3.1 Ancient Concepts of Space and Time

Concepts of Space and Time have been discussed since antiquity in both Eastern and Western philosophies and religions. Almost all religions allude to the *omnipresence* of the Creator, God: God is everywhere, all-pervading the universe in one sense or another both spatially and temporally. In the pantheistic traditions, one may find a kind of essential union between God and Universe, the latter being the substance and the former, perhaps, the soul. In the theistic traditions, which influenced the advent of modern science in the West, there is a clear separation of Creator from creation, but the physical world nonetheless reflects the attributes of God—here Space and Time may, in their essential attributes, give witness to the infinite power, presence, and personality of the Judeo–Christian–Islamic God. Insofar as physicists (Isaac Newton being the usual example) have held personal religious convictions on grounds separate from their scientific knowledge, it is natural that these should influence the metaphysical framework in which they have developed their scientific ideas.

The ancient authorities of the Western philosophical tradition had some dispute regarding the essential properties of Space. One view was of a continuously full substance or *plenum*, a philosophical conclusion based on the apparent absurdity of the idea of absolute emptiness. After all, nothing is nothing, so nothing cannot actually be, so if we follow this logic then Space cannot be empty for emptiness is an invalid property and cannot be a real state in the physical universe. The concept of plenum, or *aether*, would find a prominent place in the nineteenth-century discussions of the properties of vacuum as a medium for light waves.

The relationship between Space and the fundamental structure of matter, especially its atomicity, is especially relevant to our discussion here. In this regard, Plato, Democritus, and others of the ancient Western natural philosophers held their various views. Some ancient atomists held forth that emptiness, rather than plenum, is an essential property of our universe's Space, providing the necessary contrast to matter and the extension between material bodies in which they move. Thus, empty Space is only evident in its lack of matter, but it is really empty. The Roman philosopher Lucretius speaks of the *inane*, truly empty space that provides the opportunity for motion of the solid bodies. If Space was not truly vacant and intangible, Lucretius reasoned, then none of the solid bodies in the universe could move for there would always be something in the way!

These days, physicists shy away from the kind of semantic reasoning performed by ancient natural philosophers. Modern physics prides itself in describing Nature's properties and rules based only on physical evidence and quantitative data. But it is not clear that descriptions of Space and Time have successfully avoided an encounter with metaphysical reasoning, even in modern times. In fact, even the most recent attempts among modern physicists to consistently describe the properties of Space grapple with the distinction between these two ancient choices: Space as *inane* or Space as *plenum*. And so, for the moment at least, we will allow ourselves to exercise our thinking like natural philosophers in order to warm up to the concepts of Space and Time.

3.2 Philosophizing on Space and Time

It never hurts to begin with epistemology. We pose the following question: "How can we know anything about Space, in itself?" What is our strategy to investigate its properties and not fool ourselves by, inadvertently, studying only the properties and relations of the entities that fill Space? After all, if we are in a room, we conceptualize the room by the relationships among its boundaries and contents: the doors, walls, windows, floor, ceiling, and furniture. A geographical space is defined by the mountains, rivers, and other geographical features it contains. This includes the distances between objects and even the measuring instruments we use, within that space, to determine the distances and extensions. The important point is that any idea we have of the space around us appears to be merely a sum of the parts contained therein and nothing more. Is it not obvious, then, that the concept of Space is empty of meaning apart from the matter it contains?

Another question about Space regards its extension. What can we know about the extent of the Space that fills (or does it fill?) our universe? It is logically difficult to understand the concept that Space might have a boundary, if it has any ontological existence at all. If we travel in some direction, do we eventually reach an edge? Does not the notion of an edge, in itself, suggest that there is Space beyond the edge of Space? Another way to ask the question is this: Is the volume of Space fundamentally limited or infinite? If limited, then how, unless Space has an edge or boundary? In geometry there is indeed a way to mathematically conceptualize a bounded Space without requiring a definite boundary, but the idea seems troublesome to our common experience and common sense.

We can certainly ask similar questions about the nature of Time, and indeed modern philosophers continue to grapple with the concept. There are essentially two views of time: the tensed and the tenseless. In the former case, there is a clear distinction between events past, present, and yet to come: Only present events are actual, past events no longer exist in reality, and futurity is merely a concept of convenience. The tensed view of time makes the science fiction of time travel truly fiction, for there is no past (or future) to travel into. The tenseless view of Time has it that all times exist in some meta-present, and our human awareness is simply limited to our currently applicable time coordinate. In this sense, we are truly time traveling, just limited to a forward directionality, into the future! Just as we have formed memories only of the positions in Space that we have occupied, we have memories formed only of the times (past) that have been occupied. Any familiarity with *Minkowski spacetime* of the special theory of relativity will suggest a closer parallel between the tenseless view of Time and current thinking in the orthodoxy of physics.

Our present-day ontological perspective on Space and Time owes much to Albert Einstein's theories of relativity. As we will argue later in this chapter, Einstein's theories not only revolutionized the view of modern physics on this topic, but also reoriented the conversation within physics away from natural philosophers' preoccupation with a realist account of the basic structure of the universe, toward an effective understanding and a quasirealist interpretation. For Einstein, his successful mathematical theories overruled the ancient quest for truth about reality and delivered to us a quasirealist description of a combined Space and Time: relativistic spacetime. Before getting too deep into debating this assertion, however, it is worthwhile to recapitulate the physics issues and basic premises of scientific thinking that led Einstein to his quasirealist conclusions.

3.3   Newton's Absolute Space

Isaac Newton is well known to every student subjected to an introductory physics curriculum: He is the author of "Newton's Laws of Motion." These laws are a brief set of elegant axioms we teach our students regarding the motion and interaction of objects in everyday experience under the influence of forces. Central to these rules is the concept of inertia, the mysterious property of matter that causes it to resist changes to its state of motion. The measured masses of different objects provide a way to numerically compare how much inertia they have.

Newton's Laws of Motion arose from experience and can be stated as follows:

Newton's First Law: An object at rest will tend to remain at rest, and an object in motion will tend to retain its speed along a straight-line trajectory, unless acted upon by an external, unbalanced force.

Newton's Second Law: The effect of an external, unbalanced force acting on an object is to cause an acceleration proportional to the magnitude of the applied force and along the same direction. The magnitude of the acceleration is inversely dependent on the object's inertia.

Newton's Third Law: Every force applied by one object to another will be met with an exactly balanced counter-force of the same magnitude from the second object back onto the first.

As a starting point for his work on motion and forces, as well as his equally important labor in the field of optics, Newton had to identify the fundamental properties of physical reality that form an absolutely, ontologically real background out of which to define his rules. Physicist and philosopher Max Jammer identified the key "real" elements of Newton's natural philosophy as space, time, force, and mass. Mass is "the most essential attribute of matter" and is spatially located in mass-points. It is clear, however, that mass-points of matter are also extended in space, and so the vessel filled by mass is understood by Newton as having its own independent reality.[2]

All of Newton's laws raise important questions about the nature of Space. His first law, which is really a statement of the

---

[2]Max Jammer, *Concepts of Space: The History of Theories of Space in Physics,* 3rd Enlarged Edition (New York: Dover Publications, 1993), 99.

effects of inertia, seems to require Space to have some ontological independence from the stuff filling it. If the universe had only two objects floating in Space some arbitrary distance apart, and that separation distance was changing over time, how would we know which object was moving and which was standing still? Or perhaps both might be moving toward or apart from one another. Or consider the case where the two objects appear to be standing still with respect to one another but are in fact moving together with same velocity. These objects' intrinsic property of inertia might appreciate some clarity in this regard, so that it knows when a collision is imminent. Perhaps the background of Space can relay that information.

Reasoning that we could not properly distinguish when an object is truly at rest in that toy universe, consider an object apparently at rest in our actual universe. The law of inertia states that an applied force is required to cause it to move. But could we not produce all the necessary evidence of the beginnings of motion in that object by simply accelerating ourselves past it? Ignoring the other entities inhabiting Space around us, how could we distinguish which body is actually moving? Appealing to other bodies to help us make the determination does not help in the end, as we can inexhaustibly argue about which object in our universe to finally designate as the ultimate arbiter of the motion of all the others.

We can have sympathy, then, for Newton's conclusion that his law of inertia requires some way to ultimately specify a condition of absolute rest for a body, apart from other objects nearby that might or might not be moving past. Newton's solution to this problem was to postulate that Space, itself, is the ultimate stationary reference point of motion that allows the law of inertia to be properly maintained.

Among the properties of Space is its imperceptibility to our senses—if it has parts or features that distinguish one part of Space from another, we cannot detect them: "Absolute space, in its own nature, without regard to anything external, remains always similar and immovable."[3] Therefore, in order to usefully measure something in Space—to find one position compared to another—we must define coordinate systems that are "sensible measures" of the relative positions between objects occupying Space.[4] All of our

---

[3]Isaac Newton, *The Principia*, Great Minds Series (Amherst, N. Y.: Prometheus Books, 1995), 13.
[4]Newton, 15.

measurements, then, turn out to be relative rather than absolute, since absolute positions in Space cannot be discerned by the senses.

This also affects our measurements of motion. Newton is clear that motion is always a relative measure between bodies of which we are sensible, using relative coordinate systems:

> It is a property of rest, that bodies really at rest do rest in respect to one another. And therefore as it is possible, that in the remote regions of the fixed stars, or perhaps far beyond them, there may be some body absolutely at rest; but impossible to know, from the position of bodies to one another in our regions whether any of these do keep the same position to that remote body; it follows that absolute rest cannot be determined from the position of bodies in our regions.[5]

It is critical to see here Newton's commitment to physical realism. The problem is epistemological, rather than ontological. Our sensible observations of the universe are not observations of absolute position and absolute motion or rest. When we say an object is at rest at coordinate (x, y, z) and at time t, Newton is clear that these are effective, or mathematical, statements not to be confused with absolute measures. In principle, there is an absolute Space by which position and motion can be defined, but the relative measures are not the real things. When a measurement is made, we may refer to its outcome using the labels *space*, *place*, *motion*, and *time*, but these are by analogy, because we cannot in fact determine the actual values by comparing to the absolute Space of the universe.

Thus, "relative quantities are not the quantities themselves, whose names they bear, but those sensible measures of them."[6] To use these terms with respect to measurements of things we see around us is "purely mathematical" and to assume otherwise would be to "defile the purity of mathematical and philosophical truths, [to] confound real quantities themselves with their relations and vulgar measures."[7] Newton here appears to be anticipating the

---

[5]Newton, 15.
[6]Newton, 19.
[7]Newton, 18.

philosophy that we have labeled quasirealism and condemning it in the strongest terms![8]

Newton's second and third laws describe the effects of forces on matter, but they do not explain anything about the mechanism by which forces do their work, or the means by which they know the position at which to push or pull on an object. Some forces in Nature occur through apparent contact between objects,[9] but others are evidently pushing and pulling across a potentially vast distance. Gravitational, electrostatic, and magnetic forces act between objects that are clearly not in contact with one another.

How do these forces propagate in Space between the source and the affected object? How fast does the influence travel? Is it immediate or delayed in time? Ultimately, as expressed in the *General Scholium* he added to the second edition of his great work, *The Principia*, Newton's answer differs depending on the force. In the case of gravity, he seems to plead utilitarian ignorance, admitting,

> hitherto I have not been able to discover the cause of those properties of gravity from phænomena, and I frame no hypothesis.... To us it is enough that gravity does really exist, and act according to the laws which we have explained, and abundantly serves to account for all the motions of the celestial bodies, and of our sea.[10]

In the case of "electric and elastic" forces, as well as the effects of light and motive forces in the human body, Newton invokes "a certain most subtle Spirit which pervades and lies hid in all gross bodies; by

---

[8]Newton's conviction against quasirealism ran deep and included a religious aspect. In the same passage, he condemned those who "strain the sacred writings" by interpreting the use of words for space, time, and motion in Scripture as providing revealed insight into the structure of Creation. Even in Scripture, they should be understood as relating to everyday measurements and not absolute Space, etc. For more on this, see *The Principia: Mathematical Principles of Natural Philosophy*, by Isaac Newton, trans. I. Bernard Cohen, Julia Budenz, and Anne Miller Whitman (Berkeley: University of California Press, 1999), 36.

[9]This is, of course, just the illusion of scale: If one looks more closely at the interaction forces between two surfaces, it is evident that the force of one object pushing on another is acting across some small separation distance, such as that between the electrons and ions that form the atoms of the objects' boundaries. The concept of "touching" is, in fact, a dubious one at the microscopic scale.

[10]Newton, *The Principia*, 442–443.

the force and action of which Spirit the particles of bodies mutually attract one another at near distances, and cohere, if contiguous."[11]

It is clear Isaac Newton could say no more about the activity of forces except to admit what is obvious: they act at a distance. A force somehow knows the location of another object and exerts its influence across Space, perhaps instantaneously. After all, the planets, separated from the Sun at immense distances in our solar system, continue to pursue their orbits through what seems to be empty Space. At each instant, the Sun and the planets and other astronomical objects must somehow be aware of all the other positions and the gravitational forces being conveyed appropriately between them. Is the Space between these objects somehow facilitating this communication, and thus has some real presence, some property that allows it to transmit the gravitational message?

For both early and modern physicists, the question of what *mechanism* allows one body to act on another has always been a point of central interest. Nuclear physicist Denys Wilkinson stated the problem thus:

> There is, among physicists, an abhorrence of "action at a distance." Crudely speaking we have a revulsion from any theory that speaks of an interaction – a force – between particle A and particle B without saying how particle A becomes aware of particle B's presence. In other words we demand (admittedly on philosophical, or possibly even sentimental, grounds) that interaction should depend on communication: A and B cannot know, cannot experience a mutual force, attractive or repulsive, unless they find out via an appropriate messenger, about each other's existence.[12]

Along with the law of inertia, this philosophical discomfort with the action of forces strengthens the Newtonian case for absolute Space, even if it does so on "sentimental grounds."

Newton's convictions in this regard were debated over the course of the next 200 years by philosophers and practitioners of the emerging discipline of physical science, including such noteworthies as Leonhard Euler, Immanuel Kant, Hermann von Helmholtz, Ernst Mach, and Hendrik Lorentz. Until the mid-nineteenth century, Newton's view was generally celebrated for its religious implications

---

[11]Newton, 443.
[12]Quoted in J. H. Mulvey, ed., *The Nature of Matter*, Wolfson College Lectures 1980 (Oxford: New York: Clarendon Press; Oxford University Press, 1981), 19–20.

in the apparent connection between the attributes of an absolute Space and the familiar Biblical conception of God.[13]

In terms of the advancement of physics, however, and as Newton certainly understood, the question of the ontological reality of Space seemed to make little difference. "It is interesting to note," wrote Jammer,

> how little the actual progress of the science of mechanics was affected by general considerations concerning the nature of absolute space. Among the great French writers on mechanics, Lagrange, Laplace, and Poisson, none of them was much interested in the problem of absolute space. They all accepted the idea as a working hypothesis without worrying about its theoretical justification.... In England, by the middle of the nineteenth century, it became clear that the concept of absolute space was useless in physical practice. In that country the great success of Newtonian physics led to the paradoxical situation of the adherence to the concepts of absolute time and absolute space, on the one hand, and their absence from practical physics, on the other.[14]

In early developments of electrodynamics, it was recognized that the forces involved in electrical and magnetic phenomena behave similarly to gravitational forces and those of mechanical contact: Bodies move in response to these forces and in obedience to Newton's second law. As well, the initial assumption within the field of electrodynamics was to adopt the action-at-a-distance conviction of Isaac Newton. This assumption was strikingly supported by the commonly observed attraction of iron filings or needles to a magnet, or the deflection of a compass needle by the influence of a current-carrying conductor, first demonstrated by Hans Christian Ørsted[15] in 1820.

3.4 Lines of Force and Fields

On deeper reflection, however, the concept of forces acting instantaneously at a distance to cause motion between bodies raised

---

[13]Jammer, *Concepts of Space*, 129.
[14]Jammer, 139–140.
[15]Ørsted (1777–1851) was both a physicist and chemist whose contribution to our understanding of magnetism is memorialized by the naming of a unit after him. The oersted (Oe) is a CGS measure of magnetic field strength equal to one dyne per maxwell.

important physical questions. Newton's scientifically unsatisfactory invocation of "Spirit" as the essential force in electrodynamics was inevitably put aside as scientists in the nineteenth century examined phenomena more systematically. A useful concept was devised by Michael Faraday[16] in the form of *lines of force*, a kind of diagrammatic technique showing how a body that is the source of a force will influence other bodies in its neighborhood based on their relative positions. In the case of gravity, for example, it was simple to visualize the lines of force reaching out isotropically into Space from a massive object, growing further apart as the force of gravity weakens with distance. The lines are an illustration of the effects of the actual force and are drawn according to certain formal rules that are taught even today to undergraduates[17] in physics.

Faraday was among the generation of scientists who performed the first careful experiments on electromagnetism. Combining his own observations with those of others, including Ampère,[18] Faraday considered lines of force a solution to the question of how forces act over a distance. It was "the product of a brain that worked with visual images,"[19] as chemist and science historian Brian Silver put it.

In 1852, Faraday published a cautious but firm articulation[20] of the utility of his concept, providing an illustration of its relationship to magnetic phenomena. He reminded his readers of Ampère's experimental result showing that an electric current in a ring induces magnetism in a second ring oriented perpendicular to the current. He then discussed lines of force, illustrated by a series of sketches (see Fig. 3.1) showing the pattern acquired by iron filings for different magnet configurations. Faraday vividly demonstrated that magnetic lines of force are not always straight lines, like those

---

[16]Michael Faraday (1791–1867), a British scientist, who made many important contributions to the foundations of electromagnetism and chemistry.

[17]See any undergraduate textbook discussion on electromagnetism. To be specific, lines of electrostatic or magnetic force are now always taught in terms of a force per unit charge and related to the concept of an electric or magnetic field rather than the force, itself.

[18]André-Marie Ampère (1775–1836), a French physicist and founder of classical electrodynamics. The Système International unit of electric current is named after him.

[19]Brian L. Silver, *The Ascent of Science* (New York: Oxford University Press, 1998), 92.
[20]Michael Faraday, "LVIII. On the physical character of the lines of magnetic force," *The London, Edinburgh, and Dublin Philosophical Magazine and Journal of Science*, **3**, 20 (1852): 401–428.

associated with a gravitational source, but curved. The curvature depends on the geometry and location of the magnetic poles as well as the presence of other magnetic materials in the neighborhood.

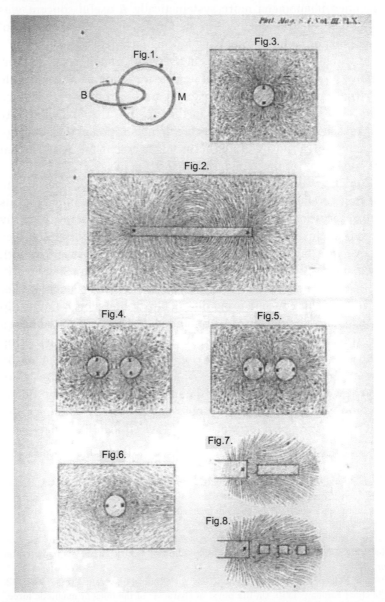

Figure 3.1 Michael Faraday's sketches of lines of force due to different magnetic sources.

What do these lines of force represent? The answer to this question was important for scientists to move beyond the starting conjecture of Newton's mysterious Spirit and add something of substance to the basic fact that a force is simply the end result of an interaction that is happening between two bodies. How is that force communicated?

Perhaps an otherwise empty Space is filled with some physical essence emanating at a finite speed from a material body, resulting in a push or pull from gravitational, electrostatic, or magnetic influences, as the case may be. James Clerk Maxwell introduced the concept of *fields* into the discussion as a way of identifying this essence that transmits the forces. Faraday's lines of force can be thought of as providing a *map* of the underlying physical field that emanates from the source object. In fact, the idea of force lines soon evolved into the concept of *field lines*, which, mathematically, are vectors related to the gradient of the field's strength.

Although the distinction may seem subtle, there are two important differences between the lines of force of Faraday and the associated fields. First, as Faraday had argued, the existence of a force is only identifiable when two bodies are involved: a source and a receiver of the force. A body itself cannot exert a force without a recipient nearby as well. A field, on the other hand, is assumed to be present in Space whether a second body is nearby or not. In fact, if the source of the field is removed, there may still be a moment in time during which the field continues to emanate away from the position where the source was located. The field exists independently of the source once it has been produced.

The second difference is evident in a distinction between gravitation and electrostatic forces on the one hand, and magnetic forces on the other. Gravitational and electrical force lines are collinear with the field vector, pushing and pulling in the direction of the field. Magnetic force lines, however, are always perpendicular to the magnetic field lines. And, whereas the gravitational and electrostatic forces act on any mass or charge, respectively, placed within the associated field, a magnetic force is only produced by the field when a charged body is in motion relative to the field lines. The fields that produce different forces have distinctive properties.

From a realist perspective, the concept of field clearly asserts that there is some actual substance with properties invisibly emanating from sources, filling intervening Space, and acting as a medium through which things move. This field responds dynamically to the presence of receptive objects, producing a force on them. Maxwell called his formulation "a dynamical theory of the electromagnetic field."[21] The strength of the magnetic force, in particular, depends on the relative velocity between the magnetic poles and the body experiencing the force, as well as the distance that separates them. Once generated by the magnetic poles, the magnetic field has an ontologically independent existence from the source and it is this field that provides the force on a charge moving through its substance. The source and receiver are now decoupled, with the field acting as a physical intermediary to affect motion in dutiful obedience to Newton's second law.

The field concept opened the door to an important consideration in the discussion of absolute Space inherited from Isaac Newton. If fields, filling the void between objects, are real substances (in some sense) that have real properties (one field being distinct from another), then we are not too far removed from accepting that Space must be a real vessel allowing for the deposition of these fields, which begin to fill the container as they emanate from their sources.

It is reasonable to assume that real fields would emanate from their sources at finite speed. It was previously known that changes in magnetic fields generate electric fields, and changes in electric fields generate magnetic fields. Maxwell famously derived a mathematical relationship capturing these facts. Maxwell's equations quantitatively describe the way electric and magnetic field strengths will change at points in space removed from the sources. Entering experimental data into his formulas, he calculated the speed of propagation for electrostatic and magnetic forces through their fields to be 310,740,000,000 mm/s, nearly identical to the velocity of light found by French physicist Hyppolyte Fizeau in 1849: 314,858,000,000 mm/s. In Maxwell's estimation, these values were in such good agreement "that we can scarcely avoid the inference

---

[21]James Clerk Maxwell, "A dynamical theory of the electromagnetic field," *Philosophical Transactions of the Royal Society of London*, **155** (1865): 459–512.

that *light consists in the transverse undulations of the same medium which is the cause of electric and magnetic phenomena*" [emphasis in original].[22]

3.5   The Aether

Once, and for all, it was a settled matter that light propagation is an electromagnetic disturbance.[23] The "electromagnetic field," Maxwell wrote in 1865, "is that part of space which contains and surrounds bodies in electric or magnetic conditions." He noted that light and heat can travel through Space even when it is "empty of all gross matter, as in the case of...so-called vacua." The concept of truly vacant Space, then, does not fit reality, and he fell back on the ancient concept of *plenum*, which he called "aether:"[24]

> We have therefore some reason to believe, from the phenomena of light and heat, that there is an aethereal medium filling space and permeating bodies, capable of being set in motion and of transmitting that motion from one part to another, and of communicating that motion to gross matter so as to heat it and affect it in various ways.[25]

This is a conceptual implication of Maxwell's spectacular mathematical work: Space is not strictly empty, but filled with a medium that has properties. Of particular importance, we are led to assign two attributes to Space, which would otherwise only make sense with respect to a material substance, namely *magnetic permeability* and *electrical permittivity*. The former is a measure of a material's ability to support a magnetic field within its substance;

[22]J. C. Maxwell, "On physical lines of force," *Philosophical Magazine*, **90**, 1 (1861): 21.
[23]At the time, the speed of light was a measured value determined from experiment. Since 1983, the speed of light has been *defined* to be exactly 299,792,458 m/s and now serves as a reference value for determining the exact length of 1 m.
[24]In the literature, we find various terminology and spelling for this medium, but we will stick to "aether" to avoid confusion with the "ether" of organic chemistry. Oddly enough, Michael Faraday can be credited with an important advance in the field of anesthesia: a published announcement in 1818 that inhalation of ether (the chemical) produces the same effects as breathing nitrous oxide [Norman A. Bergman, "Michael Faraday and his contribution to anesthesia," *Anesthesiology*, **77**, 4 (1992): 812–816]. This curious coincidence provides further reason to avoid unnecessary confusion!
[25]J. C. Maxwell, *A Dynamical Theory of the Electromagnetic Field*, 460.

the latter is a measure of a material's resistance to forming electric fields. There are definite values for these properties of the aether that are present in Maxwell's equations, determined by experiment.

This aether is assumed to be all-pervading: within atoms, inside molecules, on and between planets, in and around all stars and filling the vast intergalactic reaches of Space, everywhere.

The practical difference between Space, itself, and the aether appears to be lost in the details. Remembering Newton's conviction about absolute Space as a reference frame for measuring absolute motion, it is the aether that was now presumed to form an objective background by which an object could be seen to be truly at rest or in motion. It was determined that by examining the motion of the Earth traveling through Space, one could directly detect the presence of the aether and thus devise a way of measuring absolute motion in all other things. Earth travels around the Sun at a speed of 30 km/s and scientists in the late nineteenth century devised means of examining variations in this speed with respect to Space.

An observational phenomenon that seemed to argue for the existence of the aether was stellar aberration. This involves an apparent displacement of a star's position due to the relative motion of an object like Earth, or a telescope on Earth's surface. As a distant star emits light, the rotational and orbital motion of Earth through Space requires a periodic angular reorientation of the telescope to "catch" the star's light for continual viewing. As illustrated in Fig. 3.2, it is similar to the way an umbrella must be tilted forward when falling rain is blown sideways by the medium of the wind. The effect of stellar aberration was determined to be an angular deviation of the path of light equal to the ratio of the Earth's surface velocity to that of the speed of light.

This provided support for the conclusion that the aether is the medium of Space. Judging from our experience of other types of wave phenomena, if light is a wave then it must be an oscillation in some material medium with elastic properties of a kind that can vary from point to point along its dimensions. The fixed speed of light in the aether is not very surprising: It is consistent with the fixed speed of other waves in their respective media.

Figure 3.2 An illustration of stellar aberration. The orientation required of a telescope that is stationary with respect to the stars (bottom left) is similar to that of an umbrella when standing in rain that is falling straight down (top left). When a telescope on Earth's surface is moving relative to the stars or aether medium (bottom right), it requires a similar orientation as an umbrella in rain driven by the wind (top right).

An important study of the aether hypothesis was Michelson's[26] experiments of 1881 and 1887, the latter performed with his collaborator Edward Morley. Michelson attempted to identify any change in the speed of light due to the movement of the Earth through the aether. It is quite simple to understand the idea of Michelson's experiments by considering an analogy to two identical boats that are designed to travel in water at the same speed. If one boat must travel against a current compared to the other traveling perpendicular to the current, it is clear that the medium in which they float will have different effects on their progress. With careful

---

[26]Albert A. Michelson (1852–1931) was an American physicist. His experiments serve as a textbook example for every course in modern physics. They are: Albert A. Michelson, "The relative motion of the Earth and of the luminiferous ether," *American Journal of Science*, **128** (1881): 120–129; A. A. Michelson and E. W. Morley, "On the relative motion of the Earth and the luminiferous ether," *American Journal of Science*, **s3-34**, 203 (1887): 333–345.

observation, the presence of the medium and the speed of the current will be deducible by examining its effect on the boats as they seek to navigate a straight path away from their starting position and then make the return journey.

By a similar strategy, and using stellar aberration data as a reference, one can calculate the difference in the speed of light beginning from the same source but split along two perpendicular paths which are moving together through the stationary aether background. This experiment, along with the boat analogy, can be found in standard modern physics textbooks.[27]

As a wave, light will exhibit the effect of interference caused by the overlap of passing crests and troughs of disturbance in the medium. Michelson arranged this experiment in his laboratory using an interferometer of beam splitters and mirrors and looked for an interference pattern between the paths taken by the split light beam, indicating a difference in relative speed between the two paths. From the interference pattern, he could deduce the aether flow speed with respect to Earth, just as we can estimate the speed of a water current based on the arrival time of the two boats. There should be noticeable influences on the experiment that correlate to Earth's daily rotation on its axis, as well as seasonal variation as the planet changes direction moving around the Sun.

This was the idea. But Michelson's experiments showed that aether, if present, provides no evidence of relative motion as our planet moves through its substance. There was no evidence of any difference between the paths taken by light, as if the boats in our example are both traveling in still water, despite the other evidence of the current passing by. Since the first trials of Michelson in 1881, this experiment was repeated by himself and others into the twentieth century, confirming no change in the interference pattern.

The "null result" of Michelson's work might lead to the conclusion that either there is no medium sustaining the wave behavior of light as it travels through Space, or the medium is somehow being dragged along with Earth throughout its orbit around the Sun and is thus effectively invisible in the context of the experiment, regardless of the effects we should see due to Earth's orbit and rotation. The result was also in disagreement with the reasoning behind stellar

---

[27]See, for example, Paul Tipler and Ralph A. Llewellyn, *Modern Physics*, 5th edn. (New York: W. H. Freeman and Company, 2008), p. 8.

aberration, which was understood to show that the Earth and aether *are* in relative motion to one another.

3.6 Lorentz–FitzGerald Contraction

George Francis FitzGerald[28] commented on the work of Michelson and Morley with a letter to the editor of *Science* on May 2, 1889, and proposed a different explanation:

> I have read with much interest Messrs. Michelson and Morley's wonderfully delicate experiment attempting to decide the important question as to how far the ether is carried along by the earth. Their result seems opposed to other experiments.... I would suggest that almost the only hypothesis that can reconcile this opposition is that the length of the material bodies changes, according as they are moving through the ether or across it, by an amount depending on the square of the ratio of their velocity to that of light. We know that electric forces are affected by the motion of the electrified bodies relative to the ether, and it seems a not improbable supposition that the molecular forces are affected by the motion, and that the size of a body alters consequently.[29]

Hendrik Lorentz[30] adopted the same line of reasoning as FitzGerald in 1892 and addressed all the pertinent questions arising from stellar aberration and interferometry experiments in terms of a theory of electric and magnetic forces. Indeed, Lorentz was one of the first to visualize the atomicity of electricity in terms of elementary charges and offer a theoretical description of chemical processes in terms of electromagnetic forces between molecular electrons. In his attempt to reconcile the aether with Michelson's interferometry results, Lorentz concluded that the spacing between electrons within matter should shrink along the direction of motion through the medium. In this way, the actual length of an object will contract relative to the fixed background of absolute Space. This hypothesis became known as *Lorentz–FitzGerald contraction*.

---

[28]George Francis FitzGerald (1851–1901), an Irish physicist whose work focused on electromagnetism.

[29]George FitzGerald, "The ether and the Earth's atmosphere," *Science*, **13**, 328 (1889): 390.

[30]Hendrik Antoon Lorentz (1853–1928), a Dutch physicist, one of the founding fathers of modern physics.

The fixedness of the aether is a critical part of the recipe. The explanation of Lorentz and FitzGerald works in a straightforward way if the static aether is a reference frame for the absolute length contraction of experimental apparatuses used in the interferometry experiments. This is a realist explanation and would vindicate the Newtonian concept of absolute Space, but it is not without an important—and to some a philosophically fatal—difficulty: "If the ether as an absolute reference system could be demonstrated," wrote Jammer,

> the notion of absolute space could be saved.... Indeed, [Lorentz'] interpretation fulfilled all physical requirements...[but] shows its unsatisfactory character by the fact that it ascribes to the ether or absolute space certain definite effects which by their very assumed existence preclude any possible observation of the ether.[31]

In other words, length contraction measured relative to the fixed aether cancels out all evidence of aether's relative motion. It would be a truly invisible Space.

3.7 Special Relativity

The idea of length contraction was soon adopted in the most unlikely variation by Albert Einstein in his special theory of relativity. Einstein came on the scene of physics in the early years of the twentieth century with new and insightful thinking about the nature of Space and Time. As is well known, in 1905 he published a handful of papers that eventually brought radical changes to what physicists believed about these concepts.

Einstein's first concern was to reconcile Newton's inertial systems with electromagnetism. In a paper[32] introducing the heart of his special theory of relativity, Einstein presented his work as a clear alternative to the concept of aether, which he regarded as *überflüssig*, a *superfluous* notion. He insisted he was developing a view that would not require reference to an *absolut ruhender Raum*—an absolutely static Space—and thus Einstein set these concepts aside completely.

---

[31]Jammer, Concepts of Space, 144.
[32]Albert Einstein, "Zur Elektrodynamik Bewegter Körper," *Annalen Der Physik*, **322**, 10 (1905): 891–921.

Instead, he began with two postulates. The first was not terribly controversial: The same laws of electrodynamics and optics will be valid for all frames of reference in which the equations of motion hold good. This was in agreement with the classical *Galilean relativity* based only on the idea of mechanical forces and accelerations known in the time of Newton. Einstein's original[33] insight was to extend the postulate to electromagnetism.

His second postulate was the following: Light is always propagated in empty space with a definite speed (which we denote by the variable c, for convenience), independent of the state of motion of the emitting body. This was a more troublesome idea, and he noted it would require further elaboration as it seems incompatible with the first postulate.[34]

Einstein then considered how one treats the timing of events that occur at different locations, and he prescribed a procedure to synchronize clocks at these locations with the use of light rays. According to the postulates, these light rays should be assumed to travel at speed c through empty space independently of their direction of travel or the relative motion of light sources or observers taking measurements. In large part, the focus of his discussion is the method by which we assess and seek agreement on the time intervals and distances of even apparently simple physical scenarios, while simultaneously upholding the two postulates. The necessary consequence of the postulates is that every observer of an event has his own clocks, meter sticks, and measuring balances,

---

[33]It has long been a point of debate how much Einstein was aware of the work of others when he devised his special theory of relativity. The question may be an unfortunate consequence of the fact that Einstein's 1905 introductory paper on the topic is frustratingly sparse in its references to contemporary work on similar problems. He was aware of Michelson's experiments and at least some of Lorentz's theoretical work, although his sympathetic biographer, Abraham Pais, argues that in 1905 Einstein only knew of Lorentz's work up to 1895 [see Abraham Pais, *"Subtle Is the Lord...": The Science and the Life of Albert Einstein* (Oxford: Oxford University Press, 1983), 121].

[34]Einstein's original statement was, "In empty space light is always propagated with a definite velocity V which is independent of the state of motion of the emitting body" (trans. Anna Beck, *The Collected Papers of Albert Einstein, Volume 2: The Swiss Years: Writings, 1900–1909, (English Translation Supplement)*, p. 140, AIP Digital Library, http://einsteinpapers.press.princeton.edu/vol2-trans/154). In the intervening decades, it has been simplified to say that the speed of light is constant for all observers: an obvious misinterpretation of the original assertion.

which provide locally true readings, although they may be globally different. Measurements of location, time, and even mass[35] are variables that adjust so that the speed of light is always found to be constant regardless of relative motion.

Mathematically, Einstein's arguments led him to results identical to those of the Lorentz–FitzGerald hypothesis, including length contraction. His interpretation of these results, however, was perhaps his most significant and insightful contribution: The contraction cannot result with reference to an absolute measuring stick in the substance of the aether (or Newtonian absolute Space) but is an *apparent* contraction that occurs because of relative motion of different inertial reference frames and as a result of the constancy of the speed of light. He considered his treatment superior and more natural, as he would later write:

> On the basis of the theory of relativity the method of interpretation is incomparably more satisfactory. According to this theory there is no such thing as a 'specially favoured' (unique) co-ordinate system to occasion the introduction of the aether-idea…. Here the contraction of moving bodies follows from the two fundamental principles of the theory, without the introduction of particular hypotheses.[36]

Whether the postulates of Einstein's paper or the hypothesis of length contraction relative to a fixed aether is intrinsically "more satisfactory," the special theory of relativity provided a mathematically rigorous and consistent alternative for interpreting the experimental facts of the late nineteenth century.

The mathematician Hermann Minkowski[37] quickly recognized the merits of these new ideas on relativity and made his own important contribution to the way the theory redefined familiar concepts.

---

[35]In his concern for electrodynamics, Einstein allows that a particle's electric charge remain unchanged for different observers. It is worth pondering that a classical formulation would have a kind of equivalence between charges and masses: Electric charges generate electrostatic forces that move other particles according to Newton's second law, $F = ma$. This symmetry is broken by Einstein's treatment. It is also interesting to think that, in Nature, there exist particles with no electric charge but finite mass, but there are no particles with zero mass and yet finite charge. Yet in the special theory of relativity, charge is invariant and mass is relative.

[36]Albert Einstein, *Relativity: The Special and the General Theory*, 2nd edn. (New York: Crown Trade Paperbacks, 1961), 59.

[37]Minkowski (1864–1909): a German–Russian mathematician who had been Einstein's mathematics teacher in 1899 at what is today ETH in Zürich, Switzerland.

As will be discussed in the next chapter on Mathematical Spaces, Minkowski followed through on the mathematical implications of the theory to reinterpret the dimensionality of physical Space from a three-component coordinate system of length, width, and height, to a four-dimensional *spacetime*. In this formulation, Time and Space are mathematically united on equal ontological footing wherein they are treated as entirely equivalent relative coordinates in specifying *where* and *when* an event takes place. All four coordinates depend on the relative motion of the observer with respect to the event. Time, as now understood by physicists, loses its ontological independence and takes on a *tenseless* flavor, as discussed earlier in the chapter.

The special theory of relativity has required physicists to embrace and communicate notions about both Space and Time along lines that are at once astonishing to common experience, and at the same time absolutely necessary for Einstein's postulates, and his rejection of absolute Space, to hold true.

3.8 General Relativity

Albert Einstein's major preoccupation over the decade following his work on the special theory of relativity was to generalize the laws of physics for all observers, including those experiencing acceleration. This would be a step beyond the scope of the special theory, which applied only the context of inertial reference frames. Inertial frames would now constitute a special case from which it appears that light moves with a constant speed, c. To expand his theory to include non-inertial frames of reference, he had to revisit the ontology of Space.

Recall that Newtonian concepts prescribe a passive Space, which has an independent existence as a container in which objects exist and events occur according to the rules of physics. Within Newton's Space, objects are responsible for all their own dynamics, via their mutual interactions. Forces between particles make things happen in the Newtonian universe.

Inspired by post-Newtonian advances in electromagnetism, Einstein could envision an active Space, participating in the dynamics of the matter inside it. Maxwell's theory incorporated the very active fields of electricity and magnetism that directly influence the behavior of matter. Although the interpretation of these fields had evolved in the direction of the aether, which Einstein rejected, the field concept nonetheless encouraged him to conceive of a gravitational field as active in Space, influenced by material objects and in turn affecting the motion of bodies. He followed his thinking with respect to fields to a result where he had mathematically reconstructed the concept of Space from the inherently relative concept of the special theory, back into a kind of physical structure in which forces and fields and their influences are absorbed into the geometry of the Space, itself. Along these new lines of thinking, the notion of empty Space or vacuum was entirely abandoned.[38] In the *general theory of relativity*, Space once again takes on flesh, but not as the static reference frame of the aether. Rather, Space is a dynamic background that is itself the cause and effect of perceived forces in Nature.

The heart of Einstein's formulation of the general theory is the principle of equivalence. Roughly stated, it is the idea that the effects of gravity can be described equally well as a force pulling between material objects or as the result of an acceleration of a reference frame. From this he was led, through mathematical reasoning, to re-evaluate gravitational effects as a warping of the substance of Space due to the presence of one object so that another's motion follows an appropriately curved trajectory. In this picture, we hide the sources of forces and fields in changes to the local geometry of Space. The entire universe can be treated as a continuum of Space whose profile is ever-evolving due to the incessant presence within and movement of bodies through it. In Newtonian physics, we calculate the force (or gravitational field: force per unit mass) due to a source like the Sun on the planets orbiting nearby. In general relativity, we reduce the gravitational field, mathematically, to a change in the local spacetime geometry.

For all the equivalence of the two paradigms, however, there are real philosophical differences between Newton's and Einstein's conceptions of Space, Time, and the reality of forces. In our daily lives and professional or industrial pursuits, measurements of length, area, and volume are presumed to be simple tasks of assessing the differences in coordinates within a geometrical arrangement

---

[38]And with it the notion of empty Space, in which the speed of light is a constant, becomes a moot point in the general theory of relativity.

that every person at every position can agree on. In the Newtonian scheme, a particle subjected to a force will deviate from moving in a straight line and follow an objectively curved path. If reality follows from the mathematics of general relativity, the force/field effects are parameterized by weightage factors on the local coordinates (called *tensor matrix elements*) such that local Space receives an impromptu curvature equivalent to the Newtonian force. The weightage factors vary with location and the direction in which the length element of the particle's trajectory is measured. The consequence is that what was understood as a standard length in geometry—say 1 cm— becomes multiplied by a factor that depends on where the centimeter is, in what direction it is measured, and the state of motion of the observer(s) with respect to the centimeter.

At low speeds, Einsteinian relativity has generally been said to reduce to Newtonian–Galilean relativity, the latter being merely a special case of the more general former conception of Space and Time. From our perspective, this is a misleading assertion that obscures the profound differences between the quasirealist Einsteinian paradigm and the commitment to absolute Space engrained into the earlier Newtonian view. It may be correct to say that at low relative speeds (as $v \to 0$), the Einsteinian mathematics reduces to Newtonian equations, but it is conceptually impossible to reduce a four-dimensional Minkowski spacetime to a three-dimensional Space with a contingent Time coordinate. Even while all general relativity's multiplicative factors of tensors become unity in the Newtonian limit, the concept of the field in Einstein's picture is metaphysically distinct from the understanding of Space that emerged from the Newtonian view and the work of Lorentz and others.[39]

It is relevant to point out a fact, of which Einstein was aware, that general relativity is limited to the phenomena of gravitation and its consequences to astronomy and cosmology, and that it has not been successfully applied to what he referred to as "total field," encompassing all interactions responsible for various

---

[39]Around the same time when relativity was gaining in acceptance, distinguished mathematicians such as David Hilbert and Hermann Weyl were introducing other abstract mathematical treatments that have influenced physicists' understanding of the term "space." We address the quasirealist implications of this topic fully in Chapter 4 on Mathematical Spaces.

transformations among physical bodies. Nonetheless, with respect to gravity, the predictions of the general theory's mathematics have been impressive. No one can deny the remarkable numerical agreement of Einstein's theory with several important experiments relating to gravity,[40] including the deflection of light rays around massive bodies and the anomalous perihelion motion of the planet Mercury's orbit.

Whether the real curvature of spacetime is the responsible party for these successes, and not the gravitational force acting in an ontologically independent Space, is the important question in our concern to distinguish quasirealism from realism in the practice of physics. There is a clear mathematical usefulness to general relativity, but is there an advantage to pressing these tools to the end result of insisting on a non-Euclidean geometrization of Space? A century after the formulation of general relativity, there is still serious work to be done to provide alternative mathematical descriptions of forces and fields behaving in a way that is different from, but equivalent to, Einstein's ideas. Not all physicists have accepted relativity as an authoritative word on the nature of our universe. For example, we can point toward the very recent Relativistic Newtonian Dynamics model of Yaakov Friedman and Joseph Steiner[41] who claim to recover the numerical successes of general relativity without invoking the abstract concept of a curved spacetime. Successful alternatives to every scientific paradigm should always be pursued, even when the passing of time and familiarity have allowed generations of practitioners to simply embrace an orthodoxy—however contrary to common reason. The orthodoxy of general relativity is understandably compelling because the theory works well in a utilitarian sense, but it may be a mistake to interpret its metaphysical claims as true about reality. The practice of teaching "curved spacetime" to physics students over the past hundred years, without qualification, may yet prove to have been one long and incautious exercise in quasirealist misdirection.

---

[40]See, for example, Clifford M. Will, "The confrontation between general relativity and experiment," *Living Reviews in Relativity*, **17**, 1 (June 11, 2014): 4.

[41]Y. Friedman and J. M. Steiner, "Predicting mercury's precession using simple relativistic Newtonian dynamics," *Europhysics Letters*, **113**, 3 (2016): 39001; Y. Friedman and J. M. Steiner, "Gravitational deflection in relativistic Newtonian dynamics," *Europhysics Letters*, **117**, 5 (2017): 59001.

3.9 Concluding Remarks

Taken together, we propose that the obvious successes and aesthetics of both theories of relativity have led the physics community to an uncritical philosophical conclusion—a quasirealist acceptance that these mathematical theories are reliable guides to understanding the real structure of Space and Time in which we live. The motivation for our concern can be summarized through a few concluding quotations written along thoughtful lines of analysis by some eminent figures in the history of physics.

"The Michelson–Morley experiment," wrote Max Jammer,

> served as the starting point for the development of the theory of relativity and was interpreted by Einstein on entirely different lines [from those of Lorentz' theory], adverse to the acceptance of absolute space. It was understood that both interpretations give a complete explanation of all observations known at the beginning of the twentieth century. An *experimentum crucis* could not decide between these two theories.[42]

J. S. Bell put it more generously than Jammer: "The approach of Einstein differs from that of Lorentz in two major ways. There is a difference of philosophy, and a difference of style." The difference of philosophy is simply whether it is meaningful to talk about something as *real* (i.e., the aether or absolute Space) if it cannot be identified experimentally: "The facts of physics do not oblige us to accept one philosophy rather than the other."[43] Einstein's quasirealist view of Space is indistinguishable from Lorentz's (and Newton's) realist conception, because the mathematics work in both cases.

Einstein's quasirealist perspective is more clearly illustrated by Bell's comment about "difference in style":

> instead of inferring the experience of moving observers from known and conjectured laws of physics, Einstein starts from the *hypothesis* that the laws will look the same to all observers in uniform motion. This permits a very concise and elegant formulation of the theory, as often

---

[42]Jammer, *Concepts of Space*, 144.
[43]John S. Bell, *Speakable and Unspeakable in Quantum Mechanics*, 2nd edn., reprint, Collected Papers on Quantum Philosophy (Cambridge: Cambridge University Press, 2008), 77.

happens when one big assumption can be made to cover several less big ones. There is no intention here to make any reservation whatever about the power and precision of Einstein's approach [emphasis in original].[44]

We agree wholeheartedly with Bell's analysis of the "power and precision" of Einstein's relativity postulates but also emphasize the tension that arises in the assumptions. What Bell describes, without using the term, is characteristic of *effective* (quasi-) *theories*: a trade-off for the sake of mathematical convenience or elegance— an obscuring of the most straightforward physical interpretation so that the problem can be evaluated from a different, helpful, but artificially analogous perspective. Effective theories become food for the quasirealist philosophy of physics when physicists lose sight of the theory's roots and begin to believe and teach, without nuanced reservation, that a mathematical description is to be preferred as somehow more real than a common-sense understanding.

---

[44]Bell, 77.

Chapter 4

Mathematical Spaces

"When I use a word," Humpty Dumpty said, in rather a scornful tone, "it means just what I choose it to mean—neither more nor less."

—Lewis Carroll
Alice in Wonderland (1865)

4.1 Space and *N*-Dimensional Spaces

The physics community has undertaken the task of exploring, describing, and hypothesizing about the components of physical reality. Over time, the power and elegance that comes with adopting mathematical resources and terminology to solve physics problems may habituate a physicist to the worldview of quasirealism. This involves a seamless blending—consciously or not—of the vocabulary, tools, and shortcuts of mathematics with the entities and interactions of physical reality that mathematics is meant to effectively illustrate.

One of our main arguments in this book is that modern physics has been simultaneously enriched but also endangered by a vigorous mixing of these two worlds: mathematics and physical reality. This danger is greatest when physicists fail to remember that a non-specialist audience does not automatically understand the distinction between *methodological* quasirealism and the *ontological* characteristics of the real physical systems that mathematics usefully describes.

From Atoms to Higgs Boson: Voyages in Quasi-Spacetime
Chary Rangacharyulu and Christopher Polachic
Copyright © 2019 Jenny Stanford Publishing Pte. Ltd.
ISBN 978-981-4800-24-4 (Hardcover), 978-0-429-02765-9 (eBook)
www.jennystanford.com

In Chapter 3, we observed the historical change in physicists' collective understanding of the concept of Space. Here, we examine how this concept has been borrowed and usefully employed in physical problem-solving where the normal Space of our experience is no longer directly in view. We will illustrate a set of related cases where physicists should exercise some care in how we interpret our own ideas and communicate them to one another and an interested public.

The concept of Space is essential and elementary not only in our physicist's worldview, but for everyone: Objects and their interactions exist in what appears to us to be a three-dimensional space, and spatial parameters determine physical behavior. Because of this solid ground of familiarity, physicists and mathematicians make formal use of a variety of *N-dimensional spaces*, which are constructed by analogy to the normal Space we all think we understand. However, these new "spaces" are in fact abstract mathematical environments in which different properties of a physical system may be modeled and studied using the familiar kinds of mathematics that normally allows for objective reference to position, displacement, motion, and interaction in physical Space.

N-dimensional space simply refers to a number (N) of independent variables employed to mathematically grapple with a physical system of interest. The set or collection of these variables is given the mathematical name of a *vector*, which points to a "position" within the "space" under consideration. Each "position" represents in some way a different possible state of the system under study, in the same way that the physical location of an object in normal space can be specified by three distinct numbers ($N = 3$) relative to some starting point (for example, a distance East–West, North–South as well as an elevation when specifying a location near the surface of the earth). The vector is said to *span* the *N*-dimensional "space" it occupies, meaning that it has N components related to the N parameters needed to fully understand the physics of the system. The relationship between the real physical properties of the system and the abstract "space" of variables used by physicists is sometimes obvious to the outside observer, if provided with a clear explanation, but at other times the connection may be obscure.

In what follows, we will briefly examine some of the most common uses of these "spaces" in modern physics practice and identify the care with which we must handle them when conversing with non-specialists about these practices. Henceforth we will also drop the use of "scare-quotes" around many words that have been borrowed from the language of physical Space and trust that readers will not fall into confusion as a result. You may have noticed that we have been capitalizing the word Space whenever we mean the actual physical Space in which we move and live, and this we will continue to do to ensure clarity.

4.2   Space and Geometry

The words *space* and *geometry* do not mean exactly the same thing. However, an examination of how geometry has been used in the history of physics may lead an outside observer to think they are identical. It may be that physicists, themselves, out of everyday habit have forgotten the distinction.

It is within Space that physical entities exist and find accommodation for their extension and movement. Geometry[1] is a tool to quantify and speak clearly about Space. The concept of geometry was put on firm ground by Euclid of Alexandria (323–285 BCE), the ancient Greek mathematician who developed a formal system of principles allowing future generations to confidently work with lines, extensions, and shapes on planar surfaces and in everyday three-dimensional Space. Euclid's famous exposition of geometry was a book called *Elements*, wherein he postulated five rules that are grounded in everyday experience of Spatial measurements. The most contentious of these, the fifth postulate, is often referred to as "the parallel postulate" although here Euclid actually provided a rule for identifying two lines that are *not* geometrically parallel. It states:

> If a straight line falling on two straight lines makes the interior angles on the same side less than two right angles, the two straight lines, if produced indefinitely, meet on that side on which are the angles less than the two right angles.[2]

---

[1]The etymology of "geometry" appears to be from the Greek word *geometria*, a compound term referring to land-surveying, as indicated by its constituent parts: *geo* meaning "earth" and *metron* meaning "measure."

[2]See, for example, Herbert Meschkowski, *Noneuclidean Geometry*, trans. A. Shenitzer (New York: Academic Press Inc., 1964), 21.

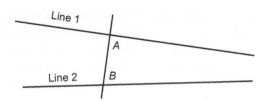

Figure 4.1   Non-parallel lines as defined in Euclid's fifth postulate of geometry.

This principle, illustrated in Fig. 4.1, involves an assumption that appears unavoidably true: If the angles A and B are each less than a right angle (that is, $A < 90°$ and $B < 90°$), then the two lines eventually meet toward the right side of the page. It is of interest to note that for the next two millennia, several mathematicians attempted to derive the parallel postulate from Euclid's first four postulates[3] with the intention of rendering the fifth redundant. They did not succeed.

Not able to prove the parallel postulate using the other four, mathematicians took on the equally noble task of disproving it and again had no success. In the end, they settled for a subversive strategy of simply devising artificial mathematical descriptions of Space in which the parallel postulate could not hold, using new rules of geometry different from the common-sense, physically inspired reasoning of Euclid. These new rules are aptly named *non-Euclidean geometries*. Within non-Euclidean geometries, designed so that the parallel postulate cannot hold, the common-sense meaning of some words such as "line" are redefined in strictly mathematical terms. Non-Euclidean lines, for example, are not *straight*—at least, not in the Euclidean, everyday sense. A non-Euclidean property of "straightness" has a mathematical definition divorced from what the word means in the context of Euclidean Space. If the property of being a line includes straightness as described by the parallel postulate, then non-Euclidean linearity or straightness simply entails a new way of using the same words. For a mathematician, this provides a refreshing new sandbox in which to play with mathematical toys; but a non-specialist—including, perhaps, a physicist—may not notice that we have moved to a different playground.

---

[3]Euclid's first four postulates:
1. Any straight line of finite length connects two points.
2. Straight lines are finite line segments made continuous.
3. Circles have a center point and radius connecting the center to equidistant points.
4. All right angles are the same.

Indeed, by formalizing non-intuitive mathematical definitions of everyday words, mathematicians have successfully devised variations on non-Euclidean geometry since the nineteenth century. The main versions that fall under this category are Riemannian[4] geometry (also known as elliptic geometry) and Gauss[5]–Bolyai[6]–Lobachevsky[7] geometry (or hyperbolic geometry). Elliptic geometry redefines the concept of a line as some entity inhabiting a space where parallel lines that obey the parallel postulate cannot exist. Hyperbolic geometry, in contrast, describes a space in which the parallel postulate is satisfied by an infinite number of different lines that pass through a single point but are all nonetheless parallel to the first line.[8]

In principle, the entities that live in the spaces of non-Euclidean geometry have little to do with the entities that inhabit the Space of our Euclidean common experience, except for a formal likeness in mathematics. The Canadian geometer Harold Coxeter wrote that the names "elliptic" or "hyperbolic" are misleading when used to describe non-Euclidean geometries: "It does not imply any direct connection with the curve [in Euclidean Space] called an ellipse, but only a far-fetched analogy. A central conic is called an ellipse or hyperbola according as it has no asymptote or two asymptotes. Analogously, a non-Euclidean plane is said to be elliptic or hyperbolic according as each of its line contains no point at infinity or two points at infinity."[9]

Euclidean planar geometry and non-Euclidean hyperbolic and elliptic geometries differ in fundamental ways, and interpreting physical reality from abstract mathematical propositions should be

---

[4]After Georg Friedrich Bernhard Riemann (1826–1866), a German mathematician known to every student of freshman mathematics for his work in integral calculus.

[5]After Carl Friedrich Gauss (1777–1855), another German mathematician and astronomer, known to every student of freshman physics for his work in electricity and magnetism. He is also connected to the bell curve in statistics.

[6]János Bolyai (1802–1860), a Hungarian mathematician.

[7]Nikolai Lobachevsky (1792–1856), a Russian mathematician. Neither Bolyai nor Lobachevsky are known to every student of mathematics or physics, despite their contributions.

[8]Several books trace the historical sequence and the roles played by these individuals in the invention of non-Euclidean geometry. See, for example, George E. Martin, *The Foundations of Geometry and the Non-Euclidean Plane* (New York: Springer, 1982).

[9]Harold S. M. Coxeter, *Introduction to Geometry*, 2nd edn, Wiley Classics Library (New York: Wiley, 1989), 94.

done with caution, and communicated to the public with even more caution.

In Euclidean geometry, the sum of the angles in a triangle is 180°, and indeed this might serve as a definition for the term "triangle." In hyperbolic geometry, the sum of a triangle is less than 180°, while it is greater than 180° in elliptic geometry. Do all three concepts of a "triangle" have the same ontological standing? Perhaps, but the physical realist might also simply assume that in the useful and ingenious mathematical expedition into the territory of non-Euclidean geometry, we have simply brought with us a few words and ideas that will be effectively re-used for different concepts. These effective concepts, such as "non-Euclidean triangle," bear some formal resemblance to the actual triangles of Space, but may be distorted by the lens of mathematics. We might think of them as *quasitriangles*, and a person who believes that quasitriangles are just as real as triangles would be holding a quasirealist interpretation. Quasitriangles may be useful, but they are not formally equivalent to triangles: They are a different shape.

To move a little deeper into this discussion, let us return to the parallel postulate that, in Euclidean Space, deals with straight lines: Their curvature is zero, and if they are parallel, the lines never meet.[10] In non-Euclidean geometries, however, an underlying curvature is introduced into the description or properties of that space, and understandably this affects the properties and behaviors of geometrical figures within that space. An often-used example of how lines would behave in a non-Euclidean Space is the particular elliptic geometry associated with the surface of a sphere. Lines drawn on the surface of a sphere have an inherent curvature to them, because they are constrained to a surface that itself possesses curvature, like the surface of the earth. Obviously, lines have no hope of being truly straight if confined in this way, and indeed a truly straight line would be either tangential to the sphere at a single point or tunnel

---

[10]Perhaps the ends of parallel straight lines meet "at infinity" even in Euclidean Space. However, the concept of an actual position at infinity in Space is of uncertain virtue. For example, see the discussion of interesting problems related to infinities discussed by George Gamow, *One Two Three...Infinity: Facts and Speculations of Science* (Bantam Books, 1961), Chapter 1. We may comfortably doubt that any *actual* measurement can find a point in Euclidean Space where two parallel lines meet.

through it and intersect the surface at two points only. As aviation companies are well aware, the shortest path between two points while constrained to a spherical surface cannot actually be a straight line, but with a proper redefinition of terms we may travel across the Atlantic Ocean by way of a "straight" flight path over Greenland. The flight path is *quasistraight*, in that it is the shortest path possible within the constraint of remaining on the actual curved surface of the Earth.[11]

The important point illustrated by this non-Euclidean example of spherical geometry is that we cannot forget where we started: A non-Euclidean space is defined as curved because we compare it to the actual straight lines of our Euclidean Space. How else could we know about the curvature without a truly straight standard of comparison? Thus, spherical geometry is a mathematical convenience of tremendous scientific, engineering, and perhaps commercial interest, but we might exercise caution in advancing a case for Space, itself, to have a non-Euclidean geometry. It is not impossible, but we need to be prepared to answer the question, "curved compared to what?" This question should be especially important for the physicist: Members of our discipline are professionally concerned with describing physical reality, and also interested in communicating concepts about reality to a listening and trusting public.

As a case in point of how even specialists appear to be confused on questions about "space" and Space, and how to define these concepts for a non-specialist audience in light of current, effective approaches, consider two statements by the same author, which were both written for a general readership to support very similar discussions about the geometry of Space. In his 2006 book *The Trouble with Physics*, physicist Lee Smolin wrote, "We are not accustomed to thinking of space as an entity with properties of its own, but it certainly is."[12] This appears to be a change in emphasis from Smolin's explanation in 1997 when he wrote, "As I hope to convince the reader in these

---

[11]Ideas about life and society constrained by the dimensionality of the Space in which one lives are imaginatively explored in Edwin Abbott Abbott, *Flatland: A Romance of Many Dimensions* (New York: Dover Publications, 1992). Abbott (1838–1926), an English school teacher and theologian, published this book in 1884, and certainly drew inspiration from ideas circulating about Space and its geometry.

[12]Lee Smolin, *The Trouble with Physics: The Rise of String Theory, the Fall of a Science, and What Comes next* (Boston: Houghton Mifflin, 2006), 41.

next chapters, space and time, like society, are in the end almost empty conceptions. They have meaning only to the extent that they stand for the complexity of the relationships between the things that happen in the world."[13] It is not that Smolin contradicts himself in what follows in these books: The arguments that follow are clear and consistent with one another, suggesting that the geometry of Space is ultimately locally variable, dependent on the gravitational field arising from the presence of matter nearby. But one might hope for less ambiguity in answer to a basic question: "Is there such a thing as Space, or not?" The web of effective ideas can overpower the clarity of even the most gifted and informed of popular science communicators.

It is the prerogative of physicists and mathematicians to adopt whatever geometry, whether Euclidean or not, that best suits the problem at hand, simplifying its solution and elucidating aspects of the system. If an entity is said to be subject to a central force like the attraction between oppositely charged electric bodies, where the attractive force depends only on the distance between the force center and the object, the physics may not be easily described in a Cartesian coordinate system of straight lines, however natural such a system might otherwise be in the absence of a central force. In this case, a more useful coordinate system (for example, polar or spherical polar coordinates) is wisely selected to solve the problem in a more fluid way, but the student of physics never loses sight of the relationship between the coordinates of this alternative system and the straight lines of the Cartesian planes. Similarly, non-Euclidean geometries have a role in helpfully but effectively describing the world, and we should always keep in mind that an entity inhabiting an inherently curved space will only know that space is curved if they compare it to a background of straight lines—that is, the truly straight lines of Euclidean Space, which is the only space of real experience.

4.3   Complex Numbers and Imaginary Planes

Another mathematical concept that has found tremendous practical, though symbolic, use in the physical sciences are *complex numbers*.

---

[13]Lee Smolin, *The Life of the Cosmos* (New York: Oxford University Press, 1997), 18.

Complex numbers were invented to solve otherwise inscrutable algebraic equations such as

$$x = \pm\sqrt{-1} \tag{4.1}$$

The solution to this kind of expression—the even root of a negative quantity—is symbolically written in terms of what are fondly called *imaginary numbers*, so-called because such values do not really correspond to anything countable in physical reality. In this case, the solution is

$$x = \pm i \tag{4.2}$$

where $i = \sqrt{-1}$ is written to remind us that there is no real solution to the problem. Now any number multiplied by this imaginary i is, itself, an imaginary number. Hiding behind the use of i, we always remember that we are trying to take the even root of a negative value, which is a futile activity in a world of countable real entities, so we proceed with caution in physically interpreting the results of our mathematical analysis.

Now consider a number line representing every possible number between negative and positive infinity.[14] It is straightforward to locate the relative positions of real numbers such as 0, –4, $-\sqrt{2}$, and π. But where do we put imaginary numbers in relation to these? Are they larger or smaller, or in between? Physically, the question does not even make sense, since i is merely a placeholder for an intractable problem. Mathematically, though, we can soldier on, taking inspiration from the way we represent an ordered pair (x, y) on a two-dimensional graph in algebra. As is well known, if we allow the x-axis to represent the real numbers, with negative infinity on the left and positive infinity on the right, then we can borrow the vertical y-axis and use it to represent the real coefficient of imaginary values. Now we have two perpendicular axes, which are independent of each other and form what we call the *complex plane*. On this abstract mathematical plane, we can plot values that have both a real and an imaginary part, which are the complex numbers.

---

[14]Which is to say, between exceedingly large values positive and negative. In physical interpretation, when a mathematical result provides the answer "infinity," it should be shunned. Roger Penrose seems to agree, saying such a result is "nonsensical" [*Fashion, Faith, and Fantasy in the New Physics of the Universe* (Princeton, New Jersey: Princeton University Press, 2016), 19]; so did Richard Feynman, who said the result would be "meaningless" [*QED: The Strange Theory of Light and Matter* (Princeton, N.J.: Princeton University Press, 1988), 127].

On this basis, it has become customary to borrow the concept of spatial rotations to understand the relationship between real and imaginary values. This invention makes the representation of complex/imaginary numbers as routine as two-dimensional geometry. As taught in elementary mathematics and physics classes, a complex number can be treated as a vector, represented by its "length" r, and some "angle of inclination" θ, with respect to a reference axis. The real part of the complex number is the vector component along the reference axis, and the imaginary part is a component along the perpendicular axis.

This is all extremely useful in application. For example, an electrical engineer can analyze an electric circuit, say that of a radio, by parameterizing electrical properties such as *resistance* (R) and *reactance* (χ), the latter of which is composed of the *capacitance* (C) and *inductance* (L) of the physical circuitry. The electrical resistance is a constant of the material, whereas the reactance varies with the frequency of the oscillating current passing through the circuit. Mathematically, the resistance can be expressed as the real component of a complex number, and the reactance as the imaginary part. Thinking of this as an ordered pair in the complex plane, illustrated in Fig. 4.2, the overall impedance is a vector of magnitude

$$Z = \sqrt{R^2 + \chi^2} \tag{4.3}$$

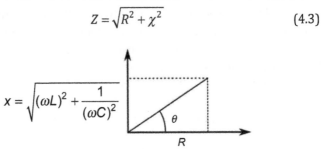

Figure 4.2 Representation of resistance (R) and reactance (χ) of an electrical circuit as a complex variable two-dimensional plane. No one misinterprets this plane to be a physical space.

This vector points at some angle θ, known as the *phase angle* in this complex space, where we have the usual trigonometric relationships

$$\cos\theta = \frac{R}{Z} \quad \text{and} \quad \sin\theta = \frac{\chi}{Z} \tag{4.4}$$

Clearly there is no physical Space where resistance and reactance are independent axes. It is by mathematical analogy that we adopt the tools that treat physical parameters that *do* have magnitudes and angular directions and employ the same language of spaces and angles as powerful instruments in the hands of professional engineers who design and build the circuitry. Surely, no one would interpret this symbolic use of a plane in complex space as something with the ontological significance as the physical Space of our common experience! To do so would be a clear example of quasirealist interpretation.

4.4 Minkowski Spacetime

We find, in fact, exactly this type of quasirealist commitment behind the scientific revolution of Einstein's relativity. The complex plane proved to be a powerful tool for reconceptualizing how distances in Space and the timing of events are measured in a universe where the speed of light, $c = 3.00 \times 10^8$ m/s, is constant in all inertial reference frames—a core postulate of the special theory of relativity. Hermann Minkowski introduced into Einstein's special relativity a kind of unification of the three classical Space dimensions with a fourth dimension of time, giving time an ontological and functional status equivalent in every (mathematical) way to length, width, and height. This unification occurs in the theory by describing the location of physical events through a four-dimensional spacetime coordinate, (x, y, z, ict), that follows from the solution to a function that requires the root of a negative number. In this coordinate, x, y, and z denote the three independent Spatial coordinates, while t is the timing of an event.

What follows from adopting this coordinate quadruplet for real events is, on the one hand, a great mathematical convenience in performing calculations for a universe with constant speed of light. From it we even deduce a four-dimensional "velocity space" and "momentum space," and following this path leads to elegant and useful simplifications when determining the energy of real systems. The mathematics is simple and fairly straightforward and was found palatable by early modern physicists weaned on the quasirealist elegance of non-Euclidean geometries and the usefulness of adopting realist terms such as *space* and *angle* to domesticate imaginary numbers within the complex plane. It was only a small conceptual leap to consider all four parts of the coordinate (x, y, z, ict) as independent of one another and having the same physical standing. After all, the fourth element, ict, being a speed–time product, has dimensionality equivalent to a length in Space.

It is now normal for us, in educating freshmen physicists, to explain Minkowski's unification along the following quasirealist pattern given in a standard textbook: "Space and time have become intertwined; we can no longer say that length and time have absolute meanings independent of the frame of reference. For this reason we refer to time and the three dimensions of space collectively as a four-dimensional entity called spacetime, and we call (x, y, z, t) together the spacetime coordinates of an event."[15]

We propose, however, that for every convenience afforded by the adoption of quasirealist thinking, other problems arise, which must be quietly swept under the rug. These arise in the awkward, obvious questions that freshman students of physics feel foolish in asking, exactly because the question appears to be too obvious. They are usually unrelated to mathematical or technical ignorance, but instead big-picture questions that demonstrate a critical imagination in a student who has not yet been trained to silently accept the orthodoxy of his or her teachers. In this case, Minkowski's symmetry between time and Space has required the student to ignore the prior fact that the *speed* of light, c, is a parameter already defined in terms of the four parameters (x, y, z, t). After all, a speed in a frame of reference can be ascertained only with respect to rulers quantifying the spatial dimensions, and clocks for the time. We instruct each observer in her frame of reference to calibrate all her tools to correctly obtain the speed of light. However, mathematics requires that the coordinates of a proper space should be completely independent of one another. As is obvious, the fourth entry of our spacetime coordinate (ict) does not meet this criterion: No matter what else it means, a speed is a measure of distance travelled over some time interval. The fourth coordinate is an imaginary value defined by observers based on their own Spatial location, as with any speed. We are thus not dealing with a set of four *independent* axes defining our space in relativity.

---

[15]Hugh D. Young, Roger A. Freedman, and A. Luis Ford, *Sears and Zemansky's University Physics: With Modern Physics*, 12th edn. (San Francisco: Pearson Addison Wesley, 2008), 1284.

What we see, then, is that familiarity with the usefulness and even indispensability of mathematical tools for simplifying problems in physics has led in this particularly important instance to a way of speaking about physical reality along quasirealist lines. Physicists grow up using vectors to represent physical properties such as an object's displacement, velocity, or momentum in Space. The same rules apply, formally, when we work with complex numbers in the complex plane. However, just because we can use a Cartesian plane to represent these artificial numbers in a useful way does not lead us to interpret the complex plane as somehow part of our real physical universe. As physicists, we are committed to scientific realism and should always be mindful of what assumptions go into our analyses. Any knowledge we gain from mathematical starting points should be filtered to ensure we are not heading down a quasirealist path of understanding Nature.

4.5   Phase Space

At the beginning of this chapter, we discussed the analogical use of physical Space to conveniently define N-dimensional "spaces" where the parameter N is the number of independent variables that are used to describe the problem at hand. The concept of a *phase space* is a powerful tool that employs this kind of mathematical analogy in applications within and beyond physics.

In the nineteenth century, the concept of the phase space was developed in the work of Ludwig Boltzmann[16] and others to uniquely describe dynamical systems[17] in terms of the various independent ways in which the system may change—what physicists refer to as the system's *degrees of freedom*. The system is said to "span an N-dimensional phase space," a phrase that, by mathematical analogy, is similar to saying an object can occupy any position in three-dimensional physical Space. Specify a value for each of the system's N parameters, and you have defined the system's current state as a point in its phase space. Specify only $N - 1$ values, and the

---

[16]Ludwig Boltzmann (1844–1906), an Austrian physicist well known for his work on statistical physics and entropy.

[17]That is, systems whose physical values at different points in space may change as time elapses.

system is located somewhere along a curved or straight line in the phase space: There is a single degree of ambiguity. Leave off another specified value, and now the system lives somewhere in a two-dimensional plane in the phase space. It is convenient to think of a system "moving" within this plane, or along its line, as time elapses and thus the value of the unspecified properties changes, just as actual motion in real Space involves a time-dependent change in one or more Spatial coordinates.

A good example of a phase space is that of a substance that may be in a gaseous, liquid, or solid state, or some combination thereof; for instance water, which may be liquid, ice, or steam depending on a few key properties. In this case, we would characterize the system by three parameters: *pressure* (P), *volume* (V), and *temperature* (T). If all three parameters are fixed in time, then water remains in an unchanging physical state. One could (and typically does) make a two-dimensional graph of the state of water depending on any two numerical values of the three key parameters P, V, and T. Such a graph, called a *phase diagram*, illustrates the phase space mathematically occupied by the water molecules. Every phase diagram of water will show a special point in its phase space where $T = 273.16$ K and $P = 611.657$ Pa. This point is called the *triple point* of water, where the molecules of a sample will be found to coexist in all three states (solid, liquid, and gas) simultaneously. At a different position in its phase space, where $T = 647$ K and $P = 22.064$ million Pa, a sample of water will show a balance between the density of steam and liquid.

Even though the phase space of water is a useful concept for analyzing its properties and response to external environmental changes over time, and the language of physical Space and motion can be naturally adopted toward this end, there is rarely any confusion that this kind of phase space is ontologically similar to the real Space of our common experience. The phase space and phase diagram concept has been usefully extended, without confusion, beyond physical systems into more abstract scientific environments. A famous example from the early twentieth century is the predator–prey model of Volterra.[18] Volterra devised equations relating to predators and prey among fish populations, predicting how these populations might change over time. His models specified how

---

[18]Vito Volterra (1860–1940), an Italian mathematical physicist whose work rigorously bridged the gap between mathematics and biology.

many predators would be too many, when they would consume all their prey and thus perish with them for lack of food. The equations translate well into a three-dimensional phase space and can be analyzed with the language of physical Space and motion. In fact, Volterra's predator–prey phase space has been applied beyond biology to economics and analysis of political systems, where independent variables of the system are identified as spatial axes relating to buying and selling of products or conflicts between parties. Needless to say, the convenience of the phase space analogy is never confused with the physical Space that inspired it.

Other examples of phases spaces abound without limit. Geophysical analysis of ore deposits can employ mathematical relationships among strain, fluid flow velocity, and energy to construct phase diagrams that are a great help for scientists and engineers to develop clear mental pictures of complex natural systems.[19] In particle physics, phase space diagrams known as Dalitz[20] plots are powerful tools for devising experiments, detector designs, and analyzing data when systems of many particles are studied in order to identify new semi-stable states of matter called *resonances*, which may inhabit the subatomic scale of the world. Studies in chaos theory employ phase space concepts and diagrams to depict fractal behavior and bifurcation of complex systems sensitive to initial conditions. The beautiful, false-color images that illustrate these phase spaces are well known from calendars and coffee table books that are purchased by the public.

The example of these chaos-inspired illustrations, however, raises an important concern for the public absorption of scientific knowledge. While specialists understand that the spaces illustrated are mere mathematical abstractions, it is doubtful that much time is taken in the notes of a calendar to educate the consumer that a picture of *strange attractors* and *fractal patterns* is an abstract visualization and not directly associated with an actual physical structure in real Space. Public exposure to fractal images is probably presumed to increase general science interest or literacy, in some unspecified way, but we fear it may unfortunately acclimatize the

---

[19]For example, Alison Ord, Mark Munro, and Bruce Hobbs, "Hydrothermal mineralising systems as chemical reactors: Wavelet analysis, multifractals and correlations," *Ore Geology Reviews*, **79** (2016): 155–179.

[20]Richard Henry Dalitz (1925–2006): an Australian–British physicist.

non-specialist public to accept quasirealist views espoused in other areas of the physical sciences.

4.6 Hilbert Space

The concept of a Hilbert[21] space is encountered by physics undergraduates as they move beyond the mere basics of the discipline and begin their studies in the formalism of quantum theory. Hilbert space extends the Euclidean Space of our experience to an arbitrary, mathematical vector space where the different dimensions represent independent parameters of a quantum mechanical system. A vector in Hilbert space thus represents the combined effect of all parameters in the mathematical formalism. Making use of the rules of algebra and calculus, this concept formally facilitates the way quantum mechanical systems have come to be interpreted physically and offers a geometric way of thinking about properties of such systems.

Hilbert space axes can be defined for any property of interest, whether Spatial position, momentum, energy, or the inherently quantum mechanical property of internal angular momentum called *intrinsic spin*. It is a powerful tool in quantum theory, and while the dimensions of a Hilbert space might include a particle's actual position in Space, it is generally understood, when first taught, to always refer to an abstract mathematical space and not a physical one.

All students, in their early quantum mechanics courses, are taught the tensor algebra of addition used in the addition of angular momentum, which relates to the internal and external rotational behavior of particles. The amount of angular momentum and the direction of the rotation can be easily characterized by vectors in a Hilbert space, and this practice is useful for predicting such seemingly unrelated phenomena as the absorption and emission of radiation.

More abstractly, Hilbert spaces allow particle physicists a straightforward tool to categorize properties of particles in a

---

[21]David Hilbert (1862–1943): a German mathematician and one of the giants of the history of mathematical physics. In 1915, Hilbert nearly beat Albert Einstein in a race to publish a final, internally consistent theory of gravity, which would become Einstein's famous general theory of relativity.

simplifying way, with new labels, as in the case of *isospin*, which mathematically treats the difference between protons and neutrons as a rotation within an electric charge space. The mathematics is tremendously convenient, but through habit of use, in all these cases, the physicist runs the risk of viewing the quantum mechanical world of Hilbert spaces through the quasirealist lens, especially in the case of isospin where a physical and real property of matter (electric charge) is not only abstracted as a rotation, but then further expressed in terms of vectors in a Hilbert "space." It is a risk, especially for students playing in the sandbox of these spaces, when more and more layers of abstraction are added to tease out results, to become too familiar with the analogical language we use. Over time this can lead to a state of affairs wherein students grow up into professional physicists who think about their work and communicate their findings to the general public using quasirealist language, unaware that the non-specialists thought they were using the language of physical realism all along.

The drift from realism to quasirealism is observed in well-publicized work done by particle physicists on the topic of the Higgs field, which we discuss more thoroughly in a separate chapter. In the present context, it is relevant to mention that the Higgs field is an outgrowth of mathematical concepts relating to an isospin Hilbert space. Theoretical physicists have, over several decades, uncovered profound insights about the workings of the subatomic structure of physical reality using quantum field theory, which is grounded on the concept of Hilbert spaces and mathematical abstractions therein. With an eye toward *symmetry* in the mathematics, physicists embraced the concept that the Space of our universe is filled with *fields*, the fundamental stuff out of which matter and energy are made. The field concept had its origin in the nineteenth-century struggle to make sense of electricity and magnetism, but it has matured into a conceptual framework that explains the properties and very existence of what otherwise are considered particles.

The Higgs field, in particular, is a universal *something* that is said to invisibly occupy all Space and somehow impart the property of mass to matter. Physicist and physics popularizer Brian Greene writes, "we are immersed in an exotic mist called the *Higgs field*" which he describes as "molasses-like," a simile meant to illustrate how its presence in Space brings about the mass of a particle.[22] Greene, of course, in writing for a popular audience, is wise to leave his explanation of the Higgs field at the level of metaphor. But at this level alone, the non-specialist reader does not have access to the vast mathematical background out of which the Higgs field is interpreted.

As with all quantum fields, the Higgs field is, first of all, a mathematical convenience. The space of the Higgs field is a weak isospin space (already an abstraction), and the field's appearance in the theory is dependent on a collection of insightful mathematical adjustments such as assigning a weak isospin quantum number to quarks and leptons to generate mathematical symmetry between these particles in the theory, while also breaking this symmetry through a carefully devised field potential that includes a negative term, which is in fact the square of a positive variable! We have a beautiful and ingenious mathematical dance out of which leaps a fully formed Higgs field that explains mass in our universe. The question for the scientific realist is whether there is a true correspondence here to an expansive physical structure filling all Space like "an exotic mist." With the Higgs field, along with other phase spaces, are we any further along in actually describing or explaining the fundamental things that make up our universe?

4.7 String Theories and Multidimensional Space

In this section, we simply wish to draw the reader's attention to one more area of thinking in modern physics wherein quasirealism appears to be at play: *string theories*. There is not just one String Theory, but many, at least five by now. All string theories have one thing in common: They envision elementary particles not as point-objects but entities of extended size, like strings. The extension of these entities overcomes certain difficulties that arise due to infinities and mathematical singularities in the solution to equations within quantum theory. Prior to special relativity, these problems could be overcome by allowing fundamental particles to be larger than points. However, with the advent of special relativity, spatially extended objects pose new problems due to the effects of length

---

[22]Brian Greene, *The Hidden Reality: Parallel Universes and the Deep Laws of the Cosmos* (New York, Vintage Books, 2011), 75.

contraction and time dilation on the way different observers in different reference frames will view the fundamental properties of the building blocks of matter. In string theories, particles are represented as a kind of one- or multidimensional string living in four-dimensional spacetime. The details of this construction are intended to solve both the quantum mechanical problem of infinities as well as the observer-dependent problems of relativity.

Along with solutions, however, string theories also create problems, in particular what are known as *Weyl* or *quantum anomalies*, which relate to lost symmetries in the treatment of gravity within relativity. In short, to solve its anomalies and rescue the important symmetries, string theorists discovered that they must mathematically rewrite the scaffolding of physical Space as 26- or 10-dimensional, depending on which type of particle is inhabiting the world. Four of these dimensions correspond to the Minkowski spacetime discussed in Chapter 3. The extra dimensions (22 or six, as the case may be) are present but unobservable to us at the scale of current experiments. These extra dimensions solve the problems.

The question of just how many dimensions are required by a string theory revolves around what kind of particle is being modeled in the theory. *Bosonic* strings require 26 dimensions to avoid anomalies, whereas *fermionic* strings only live in 10 dimensions. The extra spatial dimensions are presumed to be small in some sense. It is typically explained, by way of analogy, that they are somehow "curled up," wound so small that experiments probing down to the scale of attometers (10^{-18} m) are insensible to the presence of these spatial degrees of freedom in our universe. String theories are remarkable bodies of mathematical work and have already yielded important insights about mathematics, if not necessarily about the physics of real matter. However, insofar as these theories are held out to the public as providing substantial insight into the real structure and properties not only of the subatomic world but also of Space, itself, we find the scent of quasirealism to be pervasive in the exercise. A good litmus test of quasirealism is whether string theorists publicly elucidate a distinction between mathematically convenient entities and the properties of the real world in which they actually live.

It may be that, given sufficient time, string theorists will meet the challenge of explaining why the three Spatial dimensions of our apparent Euclidean geometry are special, while the others are too "small" to be seen down even to sub-nuclear scales. But realist skepticism should remain to keep physicists from misleading the public about what their work is accomplishing. In his thorough critique of string theories,[23] Roger Penrose has also raised important concerns about the invocation of extra physical dimensions to solve mathematical problems. After offering a detailed discussion of problems that arise for a realist understanding of string theories, he writes:

> At this stage, the reader may have become puzzled as to why string theory is being taken so seriously by such a huge fraction of the community of extremely able theoretical physicists – particularly by those directly concerned with moving forward to a deeper understanding of the underlying physics of *the world in which we actually live*.... Why [do] string theorists seem to be so unaffected by arguments against the physical plausibility of higher-dimensional space-time? [emphasis added] [24]

Penrose's own answer is that physics is a discipline that is not immune to the vagaries of *fashion*, in this case the desire of individual physicists to find astonishing insight about Reality within purely mathematical structures:

> It appears to be the case that, to many people, there is something of a romantic appeal to the idea that, hidden from direct perception, there might be a world of higher dimensionality and, moreover, that this higher dimensionality could constitute an intimate part of the actual world we inhabit![25]

His critique may be about the romanticism of professional physicists or the way their insights are interpreted by the public. In the latter case, the public cannot be expected to have any other interpretation except that which they receive from the specialists.

---

[23]Penrose, *Fashion, Faith, and Fantasy in the New Physics of the Universe* (Princeton, N. J.: Princeton University Press, 2016). This book offers a critical analysis along technical grounds, not only of string theories, but other ideas prevalent in modern physics. It involves much more than just quasirealist concerns, but we find them implicitly included in his discussion.

[24]Penrose, 82–83.

[25]Penrose, 37.

Whether or not he would agree, we see Penrose's concerns as basically those of a scientific realist confronting a quasirealist worldview among colleagues. They are the same kinds of concerns that we think are worth considering more broadly in modern physics, as discussed throughout this book, including work done on the Higgs field. It may be that physicists have begun with a borrowing of terms between normal Space and phase spaces and then compounded the abstraction more and more until some profound mathematical progress has been achieved in a theory. Such progress, however, may not be the kind that is anticipated and understood by non-specialists or students just beginning in their study of the discipline.

Physicists who communicate the results of their elegant mathematical labors to a trusting and eager public should at least be cautious about how they identify their abstract mathematical spaces with real Spaces in the real world. From the earliest days as an undergraduate, physicists develop a habit of borrowing the language of ordinary Space and geometry to also discuss and solve problems related to abstract mathematics. This borrowing is an undeniably useful practice. It has yielded many impressive results in the study of complex problems. But the history of modern physics already contains examples where physicists, by habit, have warmed up to quasirealist interpretations of their mathematical tools. If physicists are to remain scientific realists in how they think and talk about our craft, then we have a responsibility to be on guard against the encroachment of quasirealism.

Chapter 5

Mass

Bear in mind, Gentlemen, that in questions of science the authority of a thousand is not worth the humble reasoning of a single individual.

—Attributed to Galileo Galilei by François Arago
Biographies of Distinguished Scientific Men (1859)

5.1 Mass and Weight

It should come as no surprise that *mass* and *weight* are almost always used as synonyms in normal life. Technically speaking, the weight of an object is the force of gravity acting on it, wherever we find it: We quantify this weight as the object's mass multiplied by the acceleration it experiences due to local gravity. The weight of an object is a different value on the moon than on the surface of Earth, because the effects of gravity differ in each place. The weight of an object can be measured directly in the manner stated by Isaac Newton in his *Principia*: "it is always known by the quantity of an equal and contrary force just sufficient to hinder the descent of the body."[1]

From the ancient times to our modern commercial world, human transactions have required weight determinations of various commodities, whether food, building materials, or ornaments. An early standard unit of small weight measurements was the carat.

---

[1]Isaac Newton, *The Principia*, Great Mind Series (Amherst, N. Y.: Prometheus Books, 1995), 12.

From Atoms to Higgs Boson: Voyages in Quasi-Spacetime
Chary Rangacharyulu and Christopher Polachic
Copyright © 2019 Jenny Stanford Publishing Pte. Ltd.
ISBN 978-981-4800-24-4 (Hardcover), 978-0-429-02765-9 (eBook)
www.jennystanford.com

Originally, a carat referred to the weight of four carob beans, which provided a useful standard[2] of reference for many applications: Carob trees are native to southern Europe as well as Asia and were thus accessible across a large population along active trade corridors.

As experimental science evolved into the eighteenth and nineteenth centuries, the distinction between mass and weight was more carefully conceived. In Newtonian physics, mass is an inherent, invariant property of an object, unlike weight. It is independent of a particle's other properties or behaviors, including its motion. The concept of mass happens to be the first point of discussion in the *Principia*, wherein Newton defined this property as a measure of "the quantity of matter...arising from its density and bulk conjunctly."[3] By 1889, the standard kilogram was defined with reference to a physical cylinder of platinum–iridium alloy,[4] and the carat was thus redefined as a 200 mg mass, no longer requiring any reference to carob beans.

Although mass is considered a more fundamental property than weight, it is curiously inaccessible to direct measurement. It is interesting to note how our methods of measuring and assigning mass have evolved over the last two centuries. Until very recently (and only then did exceptions arise in the context of specialized experiments performed on the atomic and subatomic scale), measurements of mass have always been performed indirectly through determination of weight or applied force. We will see, however, that on smaller scales, physicists have approached the problem of mass with an evolving set of techniques to quantify it as a property of a system. In the process, the quasirealist worldview of many physicists has overtaken and redefined this property of matter, which, throughout earlier human history, had a fairly concrete, common-sense basis in everyday life.

Newton's understanding of mass was closely linked with the concept of *inertia*, which can be understood as the resistance of matter to changes in its state of motion. To measure the mass of a body, in the Newtonian system, is to provide a numerical value to the inertia of the body. Thus, while the mass of a body is not dependent

---

[2]The idea of a *standard* reference unit for measurements of matter is important even today in scientific contexts. A standard unit should be based on some object or phenomenon that has physical permanence, is easily verified, and can be reproduced as needed without unreasonable expense.

[3]Newton, *The Principia*, 9.

[4]As of 2019 it has been defined again, in terms of Planck's constant.

on its state of motion, it is understood as being directly related to the body's response to such changes when external force is applied. In fact, applying an external force is the only method of experimentally evaluating a body's inertial mass. We observe changes to its state of motion, and these changes have to do with the object's spatial displacement over time. A clear understanding of the concepts of *space* and *time* is thus required to make sense of mass measurements.

In practice, the mass of matter is deduced within the framework of Newton's first two laws of motion. The first law tells us that an object that is either at rest or moving along a straight line will continue in its state of motion unless acted upon by an unbalanced external force. In that case, Newton's second law formalized the link between mass and force through the well-known formula,

$$F = ma \tag{5.1}$$

This expression captures the observed reality that an applied force F acting on a mass m will result in an acceleration a along the same line as the force. Mass, here, is the essential, intrinsic property of the object. The force is an external imposition, which does not depend on the object's properties—either its mass or its existing state of motion.

In this Newtonian description, the inertia or mass of the object can be deduced from kinematic observables, namely its acceleration, which is simply the rate of change of the object's velocity. Any measurement of velocity requires an objective specification of an object's change in spatial position over some elapsed time, and these values should result in a unique numerical quantity for the mass of the object that is invariant with respect to one frame of reference or another. For example, we can determine the mass of an object on Earth through its weight and then use this invariant mass to unambiguously assign the object's weight on any other planet, using Newton's second law, so long as we know the gravitational force in the latter context.

It may not have been an obvious concern to Newton, or his immediate scientific descendants, but this operational definition of mass unleashes a troubled set of conceptual details that have been subject to vigorous debate. Early on, physicists and philosophers pointed to a possible distinction between various kinds of masses that a body may simultaneously possess: inertial mass, gravitational mass, electromagnetic mass, and possibly others. Different lines of reasoning have been employed by a variety of scientific and mathematical minds to define the concept of mass in a way that is rigorously non-circular, and that does not confuse the activity of merely measuring a mass with the intended goal of defining it. The most important developments pertaining to this history have been thoroughly and insightfully detailed by the late philosopher-physicist Max Jammer.[5] In the early twentieth century, the theories of relativity given to us by Albert Einstein resulted in a broad re-evaluation of the idea of mass, along with our understanding of space and time.

5.2 Mass and Relativity

In Newtonian thinking, space and time exist and evolve independently of observers or the matter occupying them, and the concept of an objectively independent force is fundamental in a measurement. Einstein's relativity dispensed with forces and gave pre-eminence to kinetic parameters such as energy, momentum, and relative velocity between inertial reference frames. From the perspective of relativity, the mass of an object can no longer be considered simply as a constant, inherent property of its matter, determined through the application of a force, but as a contingent property, dependent on the relative motion of an observer measuring the mass.

There is still, in relativity, an inherent *proper mass* or *rest mass* possessed by every quantity of matter. However, these terms are synonyms for a property that is only specified for an observer at rest with respect to the object, occupying the object's *rest frame*. An object that is in relative motion to an observer with a speed v can be shown in relativity to have a *relativistic mass*, m_r, measured by the observer as

$$m_r = \frac{m_0}{\sqrt{1 - \dfrac{v^2}{c^2}}} \tag{5.2}$$

where m_0 is the rest mass of the moving object, as determined in its own inertial reference frame.

From this formula, the relativistic mass is observer dependent, and the coexistence of these two mass concepts within the same

[5]Max Jammer, *Concepts of Mass: In Classical and Modern Physics* (Mineola, N. Y.: Dover Publications, 1997); *Concepts of Mass in Contemporary Physics and Philosophy* (Princeton, N. J.: Princeton University Press, 2000).

theory beckons the question as to whether one is "more real" than the other, or both have identical ontological status. Jammer has provided a careful exposition of different sides to this controversy and highlights "the root" of the problem: "the term 'mass' is being used in two different connotations" corresponding to "the result of two different mathematical approaches.... From the mathematical point of view both sides of the controversy can be equally well defended...and it is at this point that philosophical considerations come into play."[6]

In our view, the most important philosophical consideration is whether a physical realist should conclude that the relativistic account of mass has taught us something fundamental about the structure and properties of matter. Among physicists, the mathematical successes of the theory of relativity seem to have led to an affirmative consensus on this question. This appears evident in the very common perspective among physicists and their students that, in the low-velocity limit, relativistic mechanics reduces identically to Newtonian mechanics. This observation comes from the form that Eq. (5.2) takes when the relative velocity between object and observer goes to zero ($v = 0$): $m_r = m_0$. Under that view, Newtonian physics is merely a limiting special case of the more universally valid relativity, and Newtonian mass is identical to the rest mass in relativity.

We believe this conclusion is understandable but unjustified and only serves the purpose of solidifying quasirealist interpretations of relativity. At the first glance, it does not seem like a dramatic conceptual leap to see Newton's mass hiding in the low-velocity limit of relativity, but Jammer usefully identifies several lines of argument that have challenged the assumption on foundational grounds.[7] We will simply highlight the fact that the foundation of relativity has been built upon the concept of space and time—namely, *spacetime*— that is completely foreign to the Newtonian worldview. Relativistic spacetime assumes a four-dimensional geometry occupied by masses (where one of the four coordinates is defined in terms of an imaginary number, ict). Jammer, referencing an objection by philosopher of science Paul Feyerabend, says, "'the attempt to identify classical mass with relative [i.e., relativistic] rest mass'

---

[6]Jammer, *Concepts of Mass in Contemporary Physics and Philosophy*, 55–56.
[7]See especially, Jammer, 41–61.

cannot be made because these terms belong to incommensurable theories."[8] The terms of incommensurable theories cannot be easily compared because their view of physical reality is fundamentally different, and it should thus be clear that one can never "reduce" to another in any simple limit. The incommensurability of the two theories was alternatively analyzed and elaborated by Erik Eriksen and Kjell Vøyenli, who wrote, "The relativistic and the classical concepts of mass are intimately associated with two contradictory theories that deal with the same subject matter. Hence the classical and relativistic concepts are rival, contradictory concepts."[9]

But to a physicist, who typically prefers to view their work through the lens of scientific realism, there is an irresistible attraction to the idea that the relativistic mass is somehow equivalent to classical inertial mass. The Newtonian picture may not be able to pinpoint exactly *what* inertial mass is, in the same way that the precise nature of electric charge is elusive, but it is nonetheless an ontologically basic property that we measure (indirectly), independently of other considerations. On the one hand, the idea that inertial mass is ontologically real seems right. On the other hand, the mathematical success (or usefulness) of relativity is indisputable. At this juncture, with respect to relativistic mass, the physicist adopts the view (unwittingly) that we call quasirealism. Quasirealism rescues the "realness" of inertial mass, in keeping with Newtonian (common-sense) instincts, while celebrating the mathematical aesthetics, convenience, and power of the newer framework. In short, quasirealism unites incommensurable theories, allowing the community of physicists to believe that increasingly abstract mathematical theories are providing increasingly profound insight into *how things really are in Nature*.

Another important consequence from the special theory of relativity is the famous proportionality statement connecting mass and energy. Einstein's equation expresses the rest mass of matter, m_0, as directly related to some quantity of rest energy, E_0, with the squared speed of light, c^2, acting as a constant of proportionality:

$$E_0 = m_0 c^2 \implies m_0 = \frac{E_0}{c^2} \qquad (5.3)$$

[8]Jammer, 57.

[9]Erik Eriksen and Kjell Vøyenli, "The classical and relativistic concepts of mass," *Foundations of Physics*, **6**, 1 (1976): 123–124.

Experimental evidence of the annihilation of massive subatomic particles into radiative energy has led to an interpretation among physicists that this well-known expression should be understood as describing the *equivalence* of mass and energy. Moreover, within the axioms of special relativity, the classical assumption of conservation of energy is retained as a basic rule of Nature, while the principle of conservation of mass is found to be without basis. This provides for the nearly universal view among physicists that the energy of a system of bodies is the fundamental quantity, and the property of mass can be interpreted away as simply a variation on this ontologically more basic physical attribute. This result is reinforced by two assumptions in Einstein's mechanics: first, that measurements of relative motion can be made with respect to the measuring instrument, specified by how we conceptualize the nature of space and time (specifically, as a four-dimensional spacetime in the relativity theory); second, that relativistic mass is, indeed, a velocity-dependent parameter.

As the physics community has embraced the special theory of relativity, the concept of mass has been redefined as energy within four-dimensional imaginary spacetime. We are conceptually adrift from the concrete, common-sense physical conceptualization of mass that underpinned classical theories of physical reality. These classical ideas of mass continue, however, to resonate with physicists in the way we treat mass as a fundamental quantitative identifier of every new subatomic particle discovered in high-energy collisions. This attraction is not surprising, since the classical idea about inertial mass has been tremendously successful, right into the post-relativistic era of nuclear physics research.

5.3   Mass of Small Things

A major achievement of the seventeenth to nineteenth centuries was to define and measure the masses of atoms. It took nearly one-and-a-half centuries to achieve this feat through a variety of experimental methods that were, first of all, grounded in Newtonian assumptions about the nature of weight, mass, force, space, and time.

The chemistry of Boyle, Charles, Gay Lussac, and Avogadro (among others) established the concept that equal volumes of all gases at the same temperature and pressure contain an equal quantity of molecules. This discovery allowed for careful definitions of atomic masses, quantities that would otherwise be inaccessibly too small for experimental determination.

In the Mendeleevian scheme of classifying the chemical elements, a striking correlation exists between the presumed atomic mass of each element (collected at an electrode for a fixed electric charge) and the element's position on the periodic table. This result was obtained in a clever but simple experiment, which is still carried out in high school laboratories today: Electrodes connected to a battery induce electrolysis in chemical solution. The electric force from the battery drags the charges through the solution to the conducting electrodes. If a constant current is maintained in this process over a fixed duration of time, one measures a resultant change in the weight of the electrodes over the course of the experiment as charge is deposited. The difference in weight from start to finish can be compared with the electric charge transported in the current, providing relative measure of chemical atoms' masses from one process to another, in the form of what chemists call "molar weight."

The electron is the most elementary particle known to physics, and its mass was first determined in terms of a ratio with that particle's electric charge in the ingenious work of J. J. Thomson. Thomson's experiments on *cathode rays* established electrons as real material particles with uniquely defined properties that can be broken away from larger samples of matter. This result was followed by Millikan's determination of the electron's charge (1.6×10^{-19} C), yielding a concrete electron mass value of 9×10^{-31} kg. Millikan's result, along with the molar weights of electrolysis experiments, determines a quantity called Avogadro's number,[10] which, in turn, fixes the weights of individual molecules. This method was one of the earlier techniques used for determining the weight of a molecule.

Unavoidably, both Thomson's and Millikan's measurements of mass were indirect: They relied on canceling an electric force from

---

[10]Avogadro's Number (N_A) is the number of items (typically atoms or molecules) in the amount of a substance weighing one molar unit. It is equal to 6.022×10^{26}. For compounds or multi-atomic elements, it is read in terms of number of molecules, and for monatomic elements in terms of atoms. For example, 2.016 kg of hydrogen contains 6.022×10^{26} diatomic hydrogen molecules, while 4.0026 kg of helium contains the same number of helium atoms.

the electron's charge with an opposing force due to a magnetic field or viscous drag[11] and then applying Newton's law.

Thomson employed an electric and magnetic field tuned so that they exactly cancel in their influence on the electron's motion. Consequently, the electron hits a screen at exactly the location where it would fall without any fields present. This result is quite sensitive to the kinetic energy, or speed of the electron, which can be otherwise determined from the known value of an accelerating voltage through which the electron travels. Through the precise balancing of the electric and magnetic forces in the experimental apparatus, Thomson was able to determine the ratio of charge to mass of the electrons.

In Millikan's experiment, the combined forces of gravitation and buoyancy were balanced against an external electric field such that a charged particle drops through a liquid with constant speed. This condition allowed him to calculate the basic unit of electric charge, which, together with Thomson's ratio, permits a good determination of the electron mass.

All these procedures rely on Newtonian definitions of weight, mass, space, and time to arrive at numerical values of mass for microscopic objects. The logical train of reasoning in each classical experiment is clear and objective, and at no point are we required to invoke quasirealist redefinitions of any of these concepts. It is important to highlight this early modern success of realist physical assumptions and methods in discovering truths about microscopic entities. The experimental chemists and physicists who performed their experiments did so in the context of the unquestioned veracity of Newtonian concepts. Their common-sense assumptions, united with their careful and ingenious methods, yielded definite information about mass as an inherent property of matter.

Equally important, however, is the observation that the experimental methods of Thomson are those still employed at the heart of all modern mass measurements. This is true of measurements performed on macroscopic objects, and on mass determinations of subatomic entities. Present-day mass measurements rely on forces that are sensitive to a particle's momentum (defined in terms of its mass and velocity). There are still no experiments that directly

---

[11]Interestingly, both the magnetic and viscous force vary with the velocity of a body and fall into the category of non-inertial forces.

measure the mass, itself, of a particle. Nonetheless, the inherent property of mass is considered a critical piece of the unique identity of each species of particle forming matter.

5.4 Modern Mass Measurements of Subatomic Particles

It is observed experimentally that a charged particle[12] moving through a magnetic field will experience a deflection in its trajectory, as if a force is pushing in a direction perpendicular to both its vector of motion and the direction of the magnetic field. If the magnetic field is uniform,[13] then the particle's trajectory will take on a circular (or perhaps helical) path. The radius of the path is directly related to the momentum of the particle and inversely related to the strength of the magnetic field.

This kind of experimental scenario allows for an indirect measure of the charged particle's mass. The momentum of a particle depends on its mass, and thus the radius of its curved path will point back to the value of this inherent property. By applying a specified magnetic field strength and measuring the bending radius and velocity of the particle, its hidden mass is only a short calculation away. This kind of mass determination is an important strategy in distinguishing species of particles observed in modern experiments, and indeed in discovering particles that have never before been known to physics!

Figure 5.1 provides an illustration of this principle. It is a sketch of the different trajectories that will be followed by a proton (p), a positively charged *kaon* (K^+) and *pion* (π^+), as well as an electron (e^-) traveling through the same magnetic field (of strength 0.5 tesla, or 5000 gauss) with the same kinetic energy (0.1 GeV = 100 MeV). The vertical and horizontal axes simply identify a distance traveled in a two-dimensional plane, from a common starting point at the origin.

It is common knowledge in the particle physics community that among these species of particles, proton is the heaviest. Assuming

---

[12]That is, a particle or body that has a nonzero electric charge, such as a proton or electron.

[13]That is, of constant magnitude and direction everywhere that the particle travels.

typical mass units[14] of MeV/c^2, a proton's mass is numerically 938.3, compared to a kaon's lighter value of 493.7, pion's still lighter mass of 139.5, and the electron's comparatively small mass of 0.511. If each particle has the same kinetic energy, the heaviest and lightest particles will have the greatest and least momenta, respectively. Thus, for the particles illustrated here, the heavy protons have the largest radius of curvature to their path, bending the least as they move through space. Electrons, being by far the lightest, bend the most under the influence of the magnetic force. Note also that the electron path bends in the opposite direction to the other particles. This is because the electron has a negative electric charge, while all the others are positively charged bodies. The sign of electric charge affects the direction of the force's push.

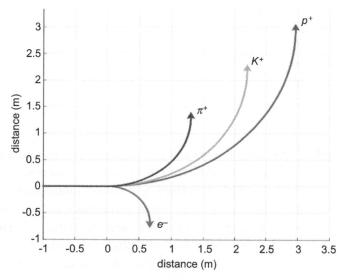

Figure 5.1   Trajectory of 0.1 GeV particles in a 0.5 T magnetic field.

Figure 5.1 is merely an illustration based on calculated position values from the equations governing the magnetic force and

---

[14]The mass units used here are MeV/c^2, which is awkward looking, but a convenient measure for very light subatomic particles. It is equivalent to 1.783×10^{-30} kg, so we can appreciate how cumbersome a kilogram would be as a mass unit for these particles. Similarly, we would not want to use the mass unit MeV/c^2 or carats to express the mass of a person. A person of, say, 50 kg would have a mass of 250,000 carats or about 2.8×10^{31} MeV/c^2.

kinematics of the particles, but very similar real particle trajectories can be directly observed in a physical laboratory. For many decades, the community of nuclear and particle physicists has observed real trajectories using a variety of detection methods. One dramatic example is the bubble chamber, in which the path of a subatomic charged particle can be visually seen as a trail of bubbles that form around gas molecules ionized by its passage.

Photographic images of particle paths in a bubble chamber, or data derived from the strategic placement of other sensors designed to detect the passage or incidence of actual subatomic particles, can be used in conjunction with precise timing instrumentation to enable physicists to determine the momentum, kinetic energy, or speed of these entities. From these data, a subatomic particle's mass can be calculated.

At high energies, charged particles also emit radiation of a form that is specific to the body's speed and the properties of the medium in which it travels.[15] Physicists have invented novel measurement techniques to exploit this phenomenon in order to further identify particles in an experimental setting.

These impressive techniques thus allow for the detection, identification, and mass measurement of subatomic particles as real entities that truly and locally pass through the experimental equipment. It is a powerful testament to realist convictions about the nature of these bodies, including a realist interpretation of the property of mass. Although these techniques require fairly sophisticated technological resources and recourse to some mathematical figuring, they are ultimately extensions of the Newtonian assumptions about space, time, motion, force, and mass.

5.5 Mass of Short-Lived Particles

These techniques allow for the observation of not only long-lived particles such as protons and electrons, but also more transient entities that may live for only a few nanoseconds before transforming into other kinds of matter and energy. Although short lived, they can

---

[15]Two types of radiation can be observed experimentally: Cherenkov and transition radiation.

travel far enough to trace out a distinguishable, directly observable path in a carefully designed detector system.

Nature appears to abound, however, in other subatomic creatures that have far shorter lifetimes and do not reveal themselves in such a tangible and straightforward manner. For these entities, with lifetimes of less than a picosecond (10^{-12} seconds), nuclear and particle physicists have devised other techniques to indirectly evaluate their properties, including mass.

When a short-lived particle ends its life in the laboratory, the universal law of conservation of energy requires that its total energy (including that which seemed to be its mass) be converted to some other form and never simply extinguished from reality. Thus, when physicists observe what appears to be evidence of new particles popping into existence they are justified in assuming that something was there beforehand to give them birth—some entity or structure that was invisible to detection, perhaps because it did not live long enough to be seen. Dying particles are understood to experience a kind of subatomic reincarnation, "decaying" but then living again as *decay products*: new particles or entities that, collectively, have conserved important properties[16] of the particle that just expired.

Identifying exactly which particle properties are thus conserved in the decay process is an extremely important subject in the history of modern physics and has resulted in some rather complicated, subtle, and abstract sections in physics textbooks. When a collection of entities is found to mysteriously appear in an experimental apparatus, their assembled properties provide an important hint at what kind of very short-lived particle just decayed. In order to determine the mass of the original particle, physicists require one more piece of information, which comes from Einstein's special relativity.

---

[16]Conservation principles exist for several classical properties of matter, known since the 1800s: energy, momentum, and electric charge. There are also more recently determined quantum conservation laws that include *parity* and *strangeness*, among others. However, these conservation laws—including that for charge—allow for the creation of an equal number of quantities of opposite sign, which still satisfy the conservation law. For example, these processes can create an equal number of particles and the corresponding antiparticles, together satisfying the conservation law.

As mentioned earlier, special relativity redefines mass as a form or expression of an object's overall energy. In so doing, the energy (E), rest mass (m_0), and momentum (p) of a particle is understood to obey the mathematical relationship

$$E^2 - p^2c^2 = m_0^2 c^4 \tag{5.4}$$

where c is the speed of light.

Experimenters can easily employ this equation to determine the mass of short-lived particles they have connected to a group of decay products, so long as they know the mass of each of these final bodies and can measure their various momenta and trajectories traveling through space. Using the sacred principles of conservation of energy and momentum, this technique has allowed for the determination that many species of subatomic entities exist even though they do not live long enough for more direct observation. This includes distinct particles that have been called the *lambda* baryon (and its relatives: the *sigma* and *omega* baryons), as well as many mesons, including those named *rho*, *omega*, and *phi*.

The energy–momentum conservation technique can also be used to identify particles that are included in decay products, or the result of the mutual annihilation of two particles that collide. Figure 5.2 shows a cleaned-up sketch[17] of a famous bubble chamber observation from the Brookhaven National Laboratory that led to the identification of the *omega baryon* subatomic particle in 1964, one of the many products that may arise when a negatively charged *kaon* collides with a proton. The lines in the image indicate physical tracks of particles within the two-dimensional plane of the liquid hydrogen bubble chamber: Solid curves are observed from bubbles in the liquid, while dashed lines are extrapolated invisible paths due to entities without electric charge that cannot cause ionization. Most paths are labeled with a symbol[18] representing the type of particle observed, and the letters A–F indicate important features in the transaction.

---

[17]Adapted from V. E. Barnes *et al.*, "Observation of a hyperon with strangeness minus three," *Physical Review Letters*, **12** (1964), 204.

[18]In the notation used by physicists to denominate their particles, the superscript indicates the electric charge of the species: "+," "–," and "0," respectively, indicate a particle that has positive, negative, or neutral charge.

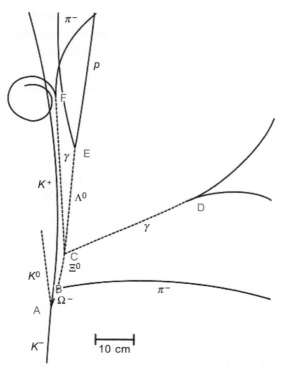

Figure 5.2 Bubble chamber tracks leading to the identification of the omega baryon.

A negative kaon (K^-) enters from the bottom of the figure and interacts at point A with a proton that is part of the background liquid of the chamber. This produces an omega baryon (Ω^-), and K^+ and K^0 mesons at the point of collision, satisfying all necessary conservation principles.[19] The two kaons live long enough to exit the chamber without further complication, but the omega flies from A to B where it decays to a *pi meson* (π^-) and a "cascade particle" labeled Ξ^0 and called the *xi baryon*. This cascade particle, being electrically neutral, leaves no visible track in the bubble chamber (hence the dashed line), but it shortly decays to a neutral *lambda* (Λ^0) as well as two *gamma rays* (γ) (which also leaves no ionization trails) at the point C. The gamma rays travel to points D and F, where they

---

[19]Notice, in particular, that overall neutral electric charge is conserved: the initial negative kaon and positive proton have zero total charge between them; the overall charge of the final three products is also zero.

each convert into *positron*[20]–electron pairs, two oppositely charged species that bend away in the magnetic field with opposite directions of curvature. Meanwhile, the lambda Λ^0 travels to point E where it decays into a proton (p) and negative *pion* (π^-).

The visible tracks in this bubble chamber image are of two pions, one proton, two electrons, and two positrons, besides the initial K⁻ meson and the positive kaon produced at point A. At points D, E, and F, we can use the visible information to reconstruct the energies, momenta, and directions of what cannot be seen, including the energy, momentum, and mass of the cascade particle as it decays at point C. The cascade particle information, along with that of the negative pion at B, is sufficient to determine the properties (including mass) of the omega baryon as it decayed at point B, or as it was produced at point A. We know the whole history of this event, and the image in Fig. 5.2 represents a spectacular success for the physics models of that day.[21]

The ability to reconstruct the masses of the various particles within this bubble chamber experiment also demonstrates the affection physicists have for a realist understanding of this property of matter. Each entity travels through the laboratory in a way that is captured by the kinematics of special relativity—if Newtonian rules are used to reconstruct the event, the predictions of the theory differ from the observation of where the particles really go in our detectors. Curiously, however, when we reconstruct energies and momenta using conservation principles we arrive back at an invariant rest mass for each particle which is often identified (in the low-velocity limit) with the Newtonian inertial mass. The relativistic factors cancel out to give us something independent of any observer's inertial reference frame, as if we have performed an inverse mathematical transformation out of special relativity back to classical physics. It is a comfortable result for scientific quasirealists.

---

[20]A positively charged antimatter partner of the electron.

[21]Particle physicists Frank Close and his co-authors described the original bubble chamber photograph of this interaction as "one of the most famous pictures in particle research, a physicist's Mona Lisa" [F. E. Close, Michael Marten, and Christine Sutton, *The Particle Odyssey: A Journey to the Heart of the Matter* (Oxford: Oxford University Press, 2004), 118].

5.6   Mass of Resonances

Not all beautiful, discrete structures seen in the detritus of particle collisions or similar experiments are the signature of a single, physical particle. They can be something else: short-lived structures that somehow fall short of earning the physical status of true particles, in an analogous way to the astronomical characterization of Pluto as something formally less than a proper planet. Particle physicists have come to call them *resonances*. The way in which masses are determined for these entities involves a series of further steps away from classical ideas about the inertial property of matter.

For a long time, the words "particle" and "resonance" were treated as synonyms. Over time, physicists began to note that not all particle/resonance signatures in an experiment should be identified on equal terms, and the words provided a convenient distinction.[22] To illustrate this development, we can examine the interaction of an electron and positron.

Figure 5.3 shows experimental data[23] from many electron–positron collisions. The horizontal axis of the diagram corresponds to the total collision energy (in units of GeV) of beam particles. The vertical axis provides a measure called the "cross section" (units of millibarns[24]), which is related to the number of events (or *yield*) detected in the apparatus corresponding to some energy. Note that the numbers on the axes are not linear, but logarithmic, so that they span a very wide range from 0.3 to 200 GeV—a factor of 1000 increase from left to right. The vertical scale allows for yield measurements differing by a factor of 1 million.

The figure includes various features that are meaningful to a physicist examining the data. In any one experiment, individual pairs of electrons and positrons will collide to produce a spray of

----

[22]Strictly speaking, one the main distinctions between a "particle" and a "resonance" is that a particle may be produced if the energy exceeds the mass of the particle, while a resonance appears at the specific energy only. One should conclude that particle physicists do not adhere to this distinction since virtual particles do not have fixed mass.

[23]C. Patrignani *et al.* (Particle Data Group), *Chinese Physics C*, **40**, 100001 (2017): 4.

[24]While the cross-sectional dimensions of millibarns are actually a measure of "area," it suffices to note that it is a measure of probability that a process occurs. Thus, a larger cross section means a higher probability for an associated outcome if we repeat the experiment under identical conditions.

new particles and background radiation, which are reconstructed through signals in an array of detectors, and appear in the graph as data points. The first item of interest is the overall downward slope of this data, approximated by a thin line. This slope shows that higher energy events are detected less frequently than those of lower energy. But this trend is sprinkled with peaks that stand out as exceptions to the rule. The detectors identify that certain entities at specific energies are reliably produced in many collisions in such large number that they are conspicuous against the overall background of radiation. These require careful analysis and are understood by physicists as signaling the detection of a particle or resonance. They are identified in the graph with different letters.

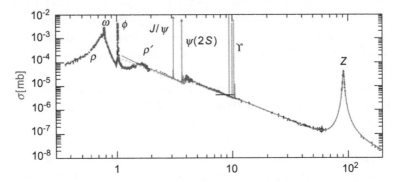

Figure 5.3 Particles and resonances appearing from electron–positron collisions. The scale is logarithmic on both axes and the horizontal axis is energy in GeV units. The vertical axis is a measure of probability (see text for details).

The letters ρ, ω, ϕ, and ρ' are particles called *mesons*. Given the relativistic equivalence of mass and energy, their masses correspond to the energies read off the horizontal scale of the graph. For example, ϕ is a meson whose data peak appears at about 1 GeV energy, and its mass can be calculated as

$$m_0 = \frac{E_0}{c^2} = 1\frac{\text{GeV}}{c^2} \tag{5.5}$$

which, through an appropriate conversion of mass units, is equivalent to 1.79×10^{-24} g. These four mesons were originally understood as true particles by physicists but today are considered composite bodies, constructed from a pair of more fundamental entities: quarks and antiquarks.

The peaks in the data lying between 3 and 4 GeV and labeled J/ψ and ψ(2s), as well as those near 10 GeV (labeled Υ and called "upsilonium"), are mesons identified as *resonances*. They have an energy corresponding to their position on the graph, but this energy is not considered to correspond to a mass as with the particles. The J/ψ and ψ(2s) structures are thought to be composite bodies, each made from a bound pair of quarks—a *charm* quark and an *anticharm* quark. These resonances are thus affectionately called "charmonium." The energy of the resonance suggests that each of these constituent quarks might have a mass corresponding to half the value: about 1.7 GeV. The Υ meson at 10 GeV, known as "bottomonium," is a composite of a *bottom* and *antibottom* pair of quarks, each with mass 5 GeV/c^2.

Here we have (approximately) identified the quark masses by simply dividing the experimental resonance energies in half. There is a long-cherished and still intact fundamental symmetry[25] of physics stipulating that a particle and its antiparticle are of identical mass values, and it is of great help in determining the masses of charm and bottom quarks in this way. But there are other quarks hiding in the features of this figure called *up*, *down*, and *strange*. Oddly enough, we do not use the same technique to determine the masses of these quarks, as will be discussed in the next section.

The broad structure appearing around 100 GeV has been identified as a signal of the famous Z boson. Along with its charged counterpart W bosons, the electrically neutral Z is deemed responsible for Nature's weak particle and nuclear interaction. The Z boson is very short lived, with a lifetime of about 10^{-26} seconds, and it decays into both lepton and hadron particles. Even so, the particle physics community has assigned to Z the status of an *elementary* boson particle, in curious contrast to the non-particle designation of the charmonium resonances, which are comparatively long lived! In general, we might expect that to identify its mass, we would simply read off the position of the peak, as before, and have our answer. But

---

[25]This symmetry is called CPT. A world made of antimatter, in which both space and time coordinates have negative signs, will look identical to the corresponding particle world with all positive coordinates. To ensure this symmetry, the masses must be equal.

the peak for the Z is quite broad,[26] and in order to arrive at a precise value a little more work is needed.

The determination of these masses includes an additional, important complication. Electrons and positrons are thought to be structureless entities and could, therefore, be expected to experience clean mutual collisions, producing well-defined energy signatures. However, very highly energetic electrons and positrons lose energy by emitting radiation. Thus, the effective collision energies are not necessarily the same as what the experimental machines are tuned for, varying from event to event. Scientists try to overcome this ambiguity through theoretical modeling: The theory of quantum electrodynamics is trusted to elucidate the influence of these effects within the experiment. The mass determination for the Z boson depends on this.

Within the context of the theory, scientists compare the results of electron–positron collisions with data from proton–antiproton collisions. Although these processes have different dependencies in the model, the consistency of results from different kinds of experiments lends credence to their ultimate conclusions about the Z mass. In 2014, after accounting for data from several experiments, the Particle Data Group[27] has assigned the mass of the Z to be 91.1876 GeV/c^2 with an error of 0.0021 in the same units—a precision of about two parts in one hundred thousand. Thus, theoretical modeling is used to refine the rough value, which, on careful inspection of the graph after the fact, does look like it may be centered over 91 GeV/c^2. Note, though, that the resolution of the actual physical instrument is supplemented by modeling, to arrive at a particle mass value, which is then used as an input back into the theory!

This raises the problem of circular reasoning, and the danger that the mass values quoted may have less correspondence to a *real* particle's mass than is portrayed to the public, or admitted to oneself and one's colleagues. Even for practicing physicists, it is difficult to carefully specify how much circular reasoning ultimately goes into this kind of relationship between modeling and experimental data

---

[26]In fact, a broad peak, like this one, also corresponds to a very short lifetime for the particle.

[27]The Particle Data Group, composed of particle physicists from around the world, prepares and publishes consensus summaries of data relating to the properties of subatomic particles and resonances, in regularly updated editions.

analysis. To quantify the actual effect would require a disciplined open mind coinciding with a simultaneous immersion in all the details of the process: a challenge at the best of times, and especially so in the competitive world of scientific research where abstract mathematical methods and numerical results can only be criticized by a close circle of participants.

Determining the mass of the W boson is more complicated still. These bosons are charged particles, and one way they decay is by emitting an electron or a muon which conserves their charge. In the process, though, a neutrino is also produced and these are elusive, escaping detection in the apparatus so that their momentum and energy are unknown in the experiment. The usual technique is to look for charged leptons of high momentum going nearly perpendicular to the axis of colliding beams. W particle identification, in the context of electron–positron collisions, also suffers the radiation effects just mentioned in the case of Z bosons. Thus, even though only one invisible neutrino escapes each decay event, we cannot reliably back-calculate the missing mass by reconstructing the decay event through energy and momentum conservation.

A further complication for finding the mass of the W is that our best theories suggest the leptons (electrons and muons) produced in their decay may not emerge from a *real* W boson, but from a *virtual* W boson. Virtual particles are understood to be entities that have physical characteristics of their real versions but are allowed to violate energy conservation principles, so long as their influence on the universe is felt for a time interval shorter than a very small value specified by Heisenberg's uncertainty principle. For example, the W boson is deemed responsible for the process in which a neutron decays into a proton, electron, and neutrino. The maximum energy involved in this kind of event is less than one GeV, and one might expect to find the mass of the W corresponding to this scale. But all such bets are off when dealing with virtual particles: The uncertainty principle provides freedom for energy conservation violation at unexpectedly large values, so long as timescales remain sufficiently short.

With this complication in mind, physicists again resort to theoretical modeling to determine an answer to the W mass question. Instead of producing W particles in electron–positron collisions, with the associated radiation problems, they can also appear in experiments using colliding protons and antiprotons. This creates new challenges, however, in that protons and antiprotons are considered to be internally complex structures, made up on the inside from quarks, antiquarks, and gluons with associated complex interaction mechanisms.

From the perspective of engineers and accelerator physicists working on these experiments, the machines are designed, built, and operated for the purpose of working with accelerated protons and antiprotons. From the perspective of particle physicists within the same research group, the focus is on the level of quarks and gluons. In their analysis, the actual collision is understood to occur among the internal quarks that travel together as a beam through the machine. Quarks and antiquarks are confined to small separation distances inside the proton and move restlessly about in this interior environment. This must be appropriately modeled in order to reliably contribute something to the study of boson production.

Layering together the results of this kind of complicated modeling to fill in the gaps of experimental data, the mass of the W boson can finally be inferred. Today we understand this value to be about 80 GeV/c^2. We recall that this particle is involved in neutron decays on energy scales of less than 1 GeV, so this is a surprising result from the perspective of energy conservation! Only with the help of Heisenberg's uncertainty principle, and the concept of virtual particles that are thought to fleet in and out of existence, can one make sense of the large scale of this mass. The virtual W boson must only survive within the neutron decay process for about 10^{-26} seconds or less in order for this to be understood as the real mass of the real (or virtual?) particle. Endowing these W and Z resonances with the property of physical reality and thus holding them responsible for neutron decay seems too convenient.

5.7 Mass of Quarks

We have already noted that the mass of some kinds of quarks—the charm and bottom varieties—can be readily determined by simply dividing the apparent mass of their associated meson by two, since mesons are said to be composed of a quark and its antiquark partner. The mass measurement of the top quark is another story, involving different techniques.

It is widely known that the top quark was first discovered in 1995 in proton–antiproton collisions in the Tevatron, a particle accelerator at the Fermi National Laboratory (Fermilab) near Chicago, Illinois. Figure 5.4 depicts physicists' model of top quark production in these collisions. A quark (q) and antiquark ($\bar{q}$), found inside the internal structure of the proton and antiproton, fuse to become a gluon (g). The gluon, in turn, becomes a top (t) and antitop ($\bar{t}$) quark pair. They decay into a new collection of particles. The top becomes a W^+ boson and bottom (b) quark, while the antitop experiences a similar decay into a W^- and antibottom ($\bar{b}$) quark. The W^+ boson finally decays into a positron (e^+) and neutrino (v). The W^- turns into a muon (μ^-) and antineutrino ($\bar{v}$). The fates of the bottom and antibottom quarks, which move away from the interaction point, are unaccounted for.

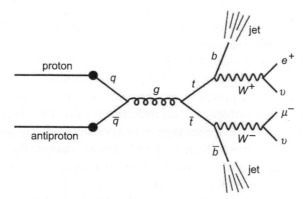

Figure 5.4 Illustration of the production of top quarks in a proton–antiproton collision.

In comparison to the identification of the W and Z boson masses, the top quark measurement has additional challenges. This particle is buried within a net of causal relationships, and its actual existence in the experimental event will be at least twice removed from the detection of real products. As a quark, it is also fractionally charged and must obey the rule of quark confinement: It cannot appear in space as a free particle in the way these other bodies can.[28] In any physical process, the top quark must be produced along with another fractional charge and these will transform to other fractional charges again. Some of them will become species of leptons and mesons, which leave a trace in detectors. They appear in the laboratory as jets of particles, but the actual top quark that created them is a virtual object whose mass determination is rather difficult.

Top quark mass was deduced from the measurements carried out at the Tevatron and by the ATLAS and CMS groups of the Large Hadron Collider at CERN in Switzerland.[29] The data from these experiments has been combined to provide a measure of the yield of top quarks as a function of beam energy across a wide range of about 1–14 TeV. This is compared with model calculations in which the top quark mass is put in as a variable parameter. A statistical match between the predictions of theory and the yield identified from the reconstruction of experimental data requires a top quark mass of 173.3 GeV/c^2.

The fact that particle physicists have arrived at a number that can be cleanly reported to the world as *the mass of the top quark*—a property of a real subatomic entity—is an impressive feat, and a testament to the technological and theoretical prowess of many people working to solve this problem over many years. It is also the result of a massively convoluted process of human negotiations, disputes over best practices in data analysis, and judgments on how to interpret ambiguous and technical results. The history and context of the particle's discovery have been impressively analyzed by science philosopher Kent Staley.[30] Staley's examination of the steps taken by a large collaborative community in arriving at this apparently simple physical result should be borne in mind. The mass value of the top quark is dressed in the context with which it is determined, involving a different set of physical assumptions from

[28]Quarks are, therefore, entities that do not have any free existence but nonetheless make their presence felt in particle physics interactions. They have similar properties to the ephemeral Cheshire Cat in Lewis Carroll's *Alice's Adventures in Wonderland*.

[29]A detailed review of the top quark properties, along with the theoretical assumptions, can be found at http://pdg.lbl.gov/2017/reviews/rpp2017-rev-top-quark.pdf. It becomes clear that the mass is derived from methods that are very different from a conventional weighing process, or bending particles in electromagnetic fields, or energy deposits of the decay products of the top quark.

[30]Kent W. Staley, *The Evidence for the Top Quark: Objectivity and Bias in Collaborative Experimentation* (Cambridge: Cambridge University Press, 2004).

the bubble chamber analyses of earlier years, or Newtonian ideas of how we determine the inertial mass of a body.

We may now introduce a final thought about the masses of quarks, along somewhat different lines. The idea that quarks have a physical property of *mass* appears to us to be seriously obscured by the fact that they cannot be understood as free entities. Furthermore, theory defines two types of quarks: *current* and *constituent* quarks. We can think of current quarks as "bare" or "naked" quarks and constituent quarks as "dressed." Accordingly, when we understand nucleons to be made up of three quarks, these are the dressed variety, and their masses are assigned a value of around 300 MeV/c^2—about one third the mass of a nucleon. It is postulated, as well, that the constituent quarks have the current quarks at their core. If this is true, then a constituent quark is actually a dressed current quark.

At the present time, it is safe to assume the heavier quarks (strange, charm, bottom, and top) are barely dressed and the constituent quarks of those species are assigned equivalent masses. However, the omega baryon, described earlier, is considered a composite of three strange quarks, just as a proton is made up of two ups and one down. This should put the mass of a constituent strange quark at about 600 MeV/c^2, rendering it most heavily dressed. Physicists appear silent about this anomaly.

The up and down quarks, which make up the major portion of normal matter, are said to be heavily dressed, since current up and down quark masses are in the range of 2–5 MeV/c^2, less than 1% of the constituent quark masses. The masses of current quarks are deduced from elaborate calculations known as lattice quantum chromodynamics. These calculations can only estimate the ratios of masses, but not the masses themselves, and must use a few experimental data as input parameters. For example, a recent calculation of nucleon masses uses the lighter quark masses and those of three mesons as input parameters and seems to produce an interpolation of the accepted proton mass to percent-level accuracy. The question that remains in all this is still the following: When we determine quark masses, are we arriving at meaningful values of a *real* inertial property of matter or something that is several steps removed from a description of real physical entities? What is the meaning of the inertial mass of an entity, which cannot exist in a free state, responding to the observable push and pull of forces, anyway?

5.8 Mass of Higgs Boson

The same kind of theoretical reasoning that provides a mass value for the top quark has also, more recently, been used to assign a mass value to the famous Higgs boson. Its mass determination is a combination of several techniques that sift the data from a colliding beam of protons, also focusing on the level of quarks and gluons.

The actual production of a Higgs particle is two steps removed from the beam that collides in the laboratory. A gluon or a quark in one colliding proton interacts with a counterpart gluon or quark, respectively, in the other proton. This interaction may produce a new quark–antiquark system or a W or Z boson, which then results in a Higgs particle either as a decay product or what is called a *Higgsstrahlung*,[31] a kind of radiating Higgs boson from the W or Z or heavy quarks. Theory tells us that Higgs bosons then decay in several possible ways. One decay mode is through the emission of two gamma rays; another, through the creation of four leptons (either four charged leptons or two charged leptons and two neutrinos); or they can also decay into a bottom and antibottom pair of quarks.

From the point of view of analyzing data, the clearest signal comes from the decay mode with the least number of detectable particles. Thus, the gamma channel was the first to be investigated and published as a discovery result.[32]

At the beginning of this analysis, there are a few points to consider carefully. First, not all gamma rays are the result of Higgs decays. They may come from several different processes such as quarks emitting radiation, known as quark *bremsstrahlung*, or even gluon radiation. Second, the detection of very high-energy gamma rays of energies up to and above 100 GeV is not as simple as registering a localized signal in one detector. This is because gamma rays produced at these high energies readily transform into charged particle–antiparticle pairs, which then emit further gamma rays, which results in a growing cascade. These cascades, known as

---

[31]*Strahlung* is a German word meaning "radiation." Electrons produce X-rays by a process known as *bremsstrahlung* (braking radiation) as they pass through matter. Unlike X-rays, which are photons with zero mass, *Higgsstrahlung* suggests a kind of radiation of massive particles.

[32]It is ironic that the two-photon decay mode of the Higgs boson was the particle physics signature for their identification since, according to the theory, photons have zero mass and do not interact with Higgs bosons.

Bhabha showers among particle and cosmic ray physicists, result is a complex cluster of tracks and signals that necessitate extremely elaborate reconstructions, involving simulations, in order to identify what particles were actually involved back up the chain of causation. Physicists have devoted a remarkable and commendable level of analytical and computational skill to sorting through enormous amounts of timing and energy data to unravel possible physical causes for the actual data they register.

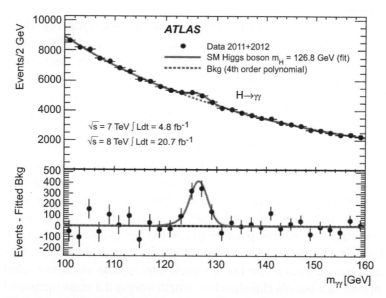

Figure 5.5 Experimental data showing a possible Higgs particle appearance from two-gamma decay events. The horizontal axis is the relativistic invariant mass of the two gamma ray photons, with units of GeV/c^2 where the parameter c^2 is left off for brevity.

The ATLAS particle physics groups at CERN investigated the two-gamma decay channel, and their results[33] are shown in Fig. 5.5. With more than 1 million events seen by the detector systems over several months, after reconstructing the invariant mass in a range of 100 to 160 GeV/c^2, as shown on the horizontal energy axis, they had to select those that fit the profile of two-gamma events.

---

[33]ATLAS Collaboration, "Plot of invariant mass distribution of diphoton candidates after all selections of the inclusive analysis for the combined 7 TeV and 8 TeV data," https://cds.cern.ch/record/1605822 (2013) (accessed April, 2018).

The data points in the top of the figure show a trend of the number of detected events decreasing smoothly as mass/energy increases, with a small local maximum at about 126 GeV comprising four data points. The dashed curve in the figure represents an estimate of all processes other than the Higgs boson, which will result in two-gamma events while also simultaneously satisfying the invariant mass in the range of interest. These are referred to as *background* events. The solid line is an estimate of the trend of actual experimental data.

Focusing on the small peak at 126 GeV mass/energy, there appear to be around 5000 background events for each of the black data points, and a total of about 1000 extra events are responsible for the excess forming the bump above the background level. Adding up these numbers, the ratio of events in the bump that exceed the level of background is about one in twenty. These are understood to provide evidence that a process matching the creation and decay of a Higgs boson has been found in the wash of data. The bottom of the figure illustrates the same peak at the same mass/energy value after the data have been cleaned up by subtracting out a number of events for each data point equal to the value of the dashed background line.

This technique of statistically analyzing bulk data from a reconstruction allows particle physicists to provide a mass for the Higgs boson equal to the position of a small data peak. It is far removed from a specific, uniquely identifiable event like what appears in a bubble chamber. It is worth asking if a mass measured in this way, based on a statistical analysis of an ensemble of events, is a real mass or a property of a feature of data that reconstructs as a mass within a certain framework of mathematical and statistical operations.

5.9   Concluding Remarks on Mass

Our treatment of mass in physics is not consistent. We assign the masses of the pi, rho, and omega to composite structures and treat them as particles. When treating the J/psi resonance, we assign the mass not to this entity as an ontologically distinct particle, but to the internal constituent quarks. For the top quark, the mass is deduced even more indirectly, through a global fit of many data sets without a direct analysis of the energies or masses of any decay products or the processes involved. We assign the masses of the Z and Higgs based upon the associated resonance energies, even though they are broad in structure and very short lived compared to other narrow structures that are not treated as particles, such the J/psi. These mass evaluations are all acquired without any direct interaction with the particle we are supposedly studying. The electron mass was never so obscure.

We should at least consider the possibility that, in these modern cases, it is an effective "mass-like" numerical value that we are assigning, by analogy, to quasireal entities. After all, what is the physical role that this property plays? After assigning mass to a subnuclear entity we may even characterize it as "virtual," to render the mass a variable parameter that can account for each physical process in which it is assigned to play a role. What, then, is the significance of mass as an intrinsic property of such a particle? It is hard to see the connection between these "masses" and the sense that mass determines a body's response to external forces or interacting partners. What purpose do the derived mass values play in the life of the fleeting W, Z, or Higgs boson, the permanently confined quark, or virtual versions of these particles that can acquire any value of mass/energy that is required, through the magic of the uncertainty principle? Is mass simply a label to distinguish one thing from another? Nuclear and particles physics have moved a long way from the Newtonian concept of mass as a measure of inertia with respect to the application of mechanical, gravitational, and other forces.

To further complicate the issue, in Einsteinian relativity we redefined the idea of mass in terms of a more fundamental energy content, with the quantitative relationship between the two involving the speed of light, c, and a detailed ontological restructuring of space, time, and the effects of moving frames of references on mass measurements. The ideas of relativity are at the heart of analyzing the particle physics experiments from which these mass determinations arise.

The concept of mass has experienced a profound evolution since the days of classical physics, and to us it appears to be a move toward physical imprecision—a technical devaluation of this otherwise necessary concept, made concrete through our experience of a body's weight. Even so, physicists continue to embrace Newtonian instincts about particle mass as an invariant and basic property of the building blocks of matter, which can be quantified as a unique identifier for every newly discovered fundamental entity. We should be alert, however, to the concern that our ideas of mass are no longer pointing us clearly to a *real* property, but rather a mathematical, statistical, or effective property. From this perspective, the continuing inclination of the physics community to speak of mass determinations as if they point to something ontologically basic is also evidence of what we identify as quasirealist physics.

Chapter 6

Quantum Physics

Science cannot solve the ultimate mystery of nature. And that is because, in the last analysis, we ourselves are a part of the mystery that we are trying to solve.

—Max Planck
Where Is Science Going? (1932)

6.1 Statistical Microphysics and Waves

Nineteenth-century physics revealed that ignorance is not a handicap. Statistical mechanics provided a means by which apparently intractable complexity could be tamed using mathematics. The philosophy of this approach is very simple. Consider a container with an enormous number of gas molecules inside it, each moving independently but bouncing off the others. The walls of the container will experience the collective influence of the individual molecules as a macroscopically observable pressure and temperature, and if the walls are moveable, a change in the enclosed volume. There is no need to know each molecule's individual kinetic parameters at any instant in time: The aggregate effect of all their interactions will provide measurable data, characteristic of the entire system. This principle was used extensively in the nineteenth century in the kinetic theory of gases developed by Maxwell and Boltzmann.

Twentieth-century physicists would inherit this approach and apply it more broadly. Modern physicists incorporate into their statistical analyses additional degrees of freedom, parameters

From Atoms to Higgs Boson: Voyages in Quasi-Spacetime
Chary Rangacharyulu and Christopher Polachic
Copyright © 2019 Jenny Stanford Publishing Pte. Ltd.
ISBN 978-981-4800-24-4 (Hardcover), 978-0-429-02765-9 (eBook)
www.jennystanford.com

that were unknown to the nineteenth-century scientists. Not all microscopic interactions are purely elastic collisions governed simply by kinematics. There are inelastic collisions that result in dissipation, absorption, and re-emission processes occurring at what Heisenberg, in his 1933 Nobel lecture, would call "unvisualizable microphysics." The description of these microphysics in the form of a statistical theory was the beginning of quantum physics, which is based on the principle that classical methods cease to be useful at a certain level of reductionist analysis. The first step was taken in modeling the interaction of light with matter.

It is well known that two incompatible views of the nature of light had competed for physicists' acceptance from the time of Newton to the twentieth century. Newton, the author of the groundbreaking book *Opticks*, exercised federal headship over those who believed light is made of *corpuscles*—particle bodies that propagate in straight lines. The competing view, which holds light to propagate as a wave phenomenon, boasted such leaders as Christiaan Huygens,[1] Thomas Young,[2] and James Clerk Maxwell.

In the system of Newtonian mechanics, a wave-like motion implies a body or substance under the influence of a periodic force. This must be so, because Newtonian matter will follow the tendency of its inertial property to propagate in a straight-line path, unless acted upon by an external force. For a free particle, waviness involves a deviation from Newton's laws.

We now know that a wave-propagating medium, such as a body of water or air, contains an immense number of molecules interacting directly with their immediate neighbors. As a tidal wave moves through water, the displacement of the waveform above and below equilibrium is fully explained by a coherent, collective movement of those molecules pushing and pulling on one another. Each molecule is subject to a net force, and we can write down the energy and momentum equations for each participant, although a complete analysis of the entire system at this level of detail will naturally escape us due to the enormous number of bodies involved. Nonetheless, we know what the appropriate expressions are, and our

[1]Christiaan Huygens (1629–1695) was a Dutch natural philosopher of many talents. Huygens discovered Titan, the largest moon of Saturn, and invented the pendulum clock in 1656. He is remembered most for his contributions to optics.

[2]Thomas Young (1773–1829) was a British physician who also made important contributions to the theories of light, elasticity of materials, and surface tension. Young's double slit experiment remains the definitive textbook justification that electromagnetic radiation propagates as a wave and not a collection of particles.

reductionist account is complete in principle. Many-body systems such as water and air have bulk wave behaviors, and we know all the details of how the microscopic reality generates the macroscopic phenomenon.

The phenomenon of electromagnetic waves absorbed and radiated from a *black body*[3] was successfully described, in microscopic mechanical terms, through a statistical treatment. The temperature dependence of the absorption and re-emission was modeled as the consequence of inelastic processes in bulk media. Electromagnetic quanta—in the form of atomic oscillators and *photons*—were invoked with discrete energies dependent on the Planck[4] constant, h, and the frequency of radiation, v:

$$E = hv \tag{6.1}$$

The true black body intensity distribution is continuous, but this model of discrete radiators provides a good approximation for the observed wavelengths of light and stimulated a paradigm shift in the thinking of physicists about microphysics. Albert Einstein recognized that, while electromagnetic radiation may propagate like waves, its interactions with matter often have the appearance of momentum and energy transfers among discrete, particle-like[5] entities: photons.

6.2 Quantum Theory of the Atom

Niels Bohr[6] took the concept of the photon one step further to develop a nuclear model of the atom. In his model, the transition of an electron between energy levels in an atom involves the

---

[3]A black body is a hypothetical object whose surface absorbs all incident electromagnetic radiation, of all wavelengths, and re-radiates it back into space. In practice, a black body's emissions can be approximated by radiation leaving a darkened cavity via a small hole.

[4]Max Planck (1858–1947) was a German physicist. Well known for the Planck constant, he first proposed the discretization of light as a statistical mathematics approach. His constant has become the cornerstone of quantum theory.

[5]We must distinguish photons from other quanta, such as electrons, in that each material object carries a minimum energy: its rest mass. Thus, each electron is of 511 keV energy as a minimum value. For a photon, whose frequencies can vary from near zero to almost infinity, we can assign the energy only when frequency is specified.

[6]Niels Henrik David Bohr (1885–1962) was a Danish physicist. Shortly after completing his doctoral degree in 1911, he spent time in the laboratory of Ernest Rutherford and was inspired to work on the theory of the atom. Bohr's influence on the foundations and later interpretation of quantum theory cannot be overstated.

emission or absorption of a photon with a frequency that provides the appropriate change in energy following Eq. (6.1). In this picture, the light behaves as a quantum of discrete energy due to the discrete nature of atomic energy levels. In general, however, the electromagnetic spectrum is a continuous one with energies (and frequencies) extending from zero to infinity.

Bohr's model was a source of consternation to the contemporary physics community. It was built on two awkward postulates: first, that electrons are restricted to allowed orbits; second, and consequently, that they transition between orbits without traversing the intervening space, through the mysterious process of *quantum jumps*. Surprisingly, even today, our classroom teaching hangs on to this picture due to its simplicity and useful extension into electronic shell structures.

Soon, Max Born[7], Pascual Jordan,[8] and Werner Heisenberg contributed to this quantum model of mechanics. The disconcerting quantum jumps were explained away by re-envisioning the localized planetary electron as a distributed cloud surrounding the atomic nucleus. Then they turned to *Fourier expansion*, a common mathematical technique used by engineers in the analysis of waveforms. Simply speaking, a continuous function can be written as the sum of a series of discrete Fourier components, whose amplitudes are connected to the contributions of different wave frequencies. Interestingly, when the electron cloud size is constrained to atomic dimensions, the mathematical results are consistent with the phenomena from the spectral analysis of light emitted from the hydrogen atom known experimentally since the nineteenth century as the Balmer series. This connection to experiment was a major achievement of the model.

The technique of Fourier analysis was Heisenberg's path to the development of a quantum theory, which he ultimately expressed in the mathematical form of matrices. A different path was taken by Erwin Schrödinger, both conceptually and mathematically. He was motivated by the work of Louis de Broglie, whose doctoral thesis

---

[7]Max Born (1882–1970) was a German physicist who won the Nobel Prize in physics in 1954 for his interpretation of the quantum mechanical wave function as a probability distribution of the physical system.
[8]Ernst Pascual Jordan (1902–1980) was a German physicist who helped to develop quantum field theory.

argued for the wave-like properties of material entities, just as light was now understood to simultaneously display both wave-like and particle-like behavior. Schrödinger argued that particle-like motion along a straight-line trajectory must be an idealization. His reasoning was sensible enough: Phenomena such as diffraction or interference of light are seen only when the apertures or opaque disks along the path of light are comparable in size to the wavelength of the light. The motion of submicroscopic entities such as electrons can be discerned only when the measuring apparatus is of comparable size to them as well, so it might be appropriate to see this as a clue that electrons are also wavy, which would be consistent with de Broglie's hypothesis. Around that time, electron diffraction was observed in the experiments of Davisson and Germer.[9] Schrödinger wrote a wave equation for the motion of a particle in a homogeneous potential and fully recovered Bohr's solutions without invoking the unfortunate stationary orbits. This was a great success, although it must be clearly noted that the particle in this model is no longer the free entity of Newtonian mechanics. It is beginning to look like a quasi-Newtonian particle.

At this stage, Schrödinger's work reminds us of the *Hamiltonian Principle*, which was, perhaps, the first attempt to unify *geometric* and *wave optics*. Geometric optics models light propagation as *rays*, which are arrows pointing in the direction of propagation of the wave fronts of light in wave optics. A ray mimics the behavior of a wave front when it encounters material media, changing the direction in which it points just as the wave front experiences bending through refraction due to the change in the index of refraction across a boundary region.

This principle was conceived by William Rowan Hamilton[10] to describe the motion of a light ray in a medium of varying refractive index. As is taught in beginner's optics courses, in this situation the

---

[9]C. J. Davisson and L. H. Germer, "Reflection of electrons by a crystal of nickel," *Proceedings of the National Academy of Sciences of the United States of America*, **14** (4) (1928): 317–322.

[10]William Rowan Hamilton (1805–1865) was an Irish physicist and mathematician whose contributions to classical physics were extensive, including a mathematical formalism for mechanics that found great application in the development of quantum theory. He also devised an extension of the complex numbers in the form of *quaternions*.

light ray is continually bent.[11] That is, the ray is not moving in a field-free region but is subject to an external force. This is, perhaps, the beginning of merging entities and media into one complex, effective description of the behavior of entities themselves. Schrödinger's wave equation can be derived from Hamilton's principle by imposing suitable boundary conditions.

It is interesting to note that Schrödinger's work in early quantum theory was an attempt to describe physical reality to the extent that it can be measured or observed in a laboratory. He eventually changed his mind about the veracity of quantum mechanics and drifted away from "orthodox" physics to pursue other questions relating to biology, life, and consciousness. His famous Schrödinger Cat paradox, in which a too-literal interpretation of the formalism of quantum theory leads to a physical cat being simultaneously dead and alive in a box, was designed to show the ridiculous outcomes of mainstream interpretations of quantum mechanical reasoning.

6.3 What Evolves in Quantum Theory?

It turns out that Schrödinger's results can be combined with Heisenberg's matrix mechanics to arrive at a unified treatment of quantum physics. However, there are three equivalent ways in which the mathematical formalism of quantum theory can be understood when we want to relate its components back to physical reality. In textbooks, these are known as "pictures" or "representations" of the theory and are called *the Heisenberg picture*, *the Schrödinger picture*, and the *interaction picture* (the last due to Paul Dirac).

It is a fundamental axiom in quantum physics that physical processes can be divided into two components: constituents, represented by *wave functions*; and *operators*, represented by mathematical operations on the wave functions,[12] which physically correspond to measurements on the constituents. If a system evolves in time, changing its properties or undergoing transitions, one may rightly inquire: Who or what is the true cause behind the observable effects? The three pictures provide different answers to

---

[11]It is also the principle of light transport by Gradient Index (GRIN) optical fibers used in optical communications.

[12]In quantum field theories, the two components are *fields* and *propagators*.

this question, and this should be troubling from a physical realist point of view.

According to the Heisenberg picture, the operators are the entities that change over the course of time, while the wave functions do not evolve. Thus, transitions from one state to another are found within the temporal dependence of the operators. In the Schrödinger picture, it is the wave functions that carry temporal dependence and operators are time independent. In the interaction picture, both the operators and wave functions share in the temporal evolution of a system.

In itself, three different ways of assigning time dependence in the equations are not obviously troublesome. Assuming that the formal elements of quantum theory correspond identically (or nearly so) to aspects of the physical systems they model, we can surely call on physical reality, itself, to adjudicate between these different representations and inform us what physical aspect of a quantum scale system truly evolves in time, and what remains unchanged. Unfortunately, there are no observable consequences that distinguish between these three different pictures and we are left in an ambiguous position. On the one hand, these three pictures are mathematically equivalent—interchangeable, in fact. On the other hand, they offer up fundamentally different physical interpretations of what is happening in nature. These two facts, together, are an important clue that the quantum theory does not provide us with direct knowledge of physical reality but is only an effective mathematical description. If quantum mechanical formulations reveal physical reality in its finest, reductionist detail, should we not expect to be able to discern the most correct of these three options?

Needless to say, succeeding generations of physicists have simply sidestepped the foundational questions about physical reality that arise in the mathematical descriptions of quantum theory, of which much more could be said. The most interesting paradoxes of quantum mechanics are well rehearsed in uncountable books at all levels of sophistication and trustworthiness, and here we will avoid repeating what can be read elsewhere, although in part of the next chapter we will explain in more detail how quasirealist thinking about quantum theory yields unfortunate consequences for our understanding of the electron.

For now, our emphasis will be simply this: In light of the rest of our arguments in this book, we encourage our colleagues in the physics community to consider the case that quantum theory may be best understood as an effective theory, of the same flavor as the statistical physics of the nineteenth century. Indeed, the simplest ensemble interpretations of the theory plausibly suggest as much, and the historical thread in the theory's development supports it as well. It is true, as well, that quantum theory is one of the foundations on which modern particle physics is constructed, and many of the quasirealist complications in that discipline, outlined in the following chapters of this book, may well (in part, at least) be traced back to this source.

As with thermodynamics, quantum theory is extraordinarily useful and has been applied with tremendous ingenuity by solid-state physicists, engineers, and others who are able to harness its guidance for technological innovation. We suggest, however, that it is a mistake to believe that quantum theory provides direct insight into the real structure and rules of the microscopic world. To admit this to the public, however, would require tremendous humility at this juncture in the history of physics: Over successive generations, we have built our reputation as the unassailable gatekeepers of physical reality in large part on the inherent paradox and oddity of the quantum world, secrets that mere mortals cannot fathom without the guidance of those of us who have been initiated into the mystery religion of physics through a successful thesis defense. To let go of that privileged status, admitting that our microscopic theory does not necessarily represent a true end to the ancient quest of physical reductionism would take great courage.

Chapter 7

When Is an Atom?[1]

*Science has explained nothing; ...the more we know the more fantastic
the world becomes and the profounder the surrounding darkness....*

—Alduous Huxley
Along the Road: Notes and Essays of a Tourist (1925)

7.1 The Classical Atom

During the recent decades, it has become clear that there is no
unique, objective physical characteristic that allows us to decide
when we should and should not call an entity a *fundamental
particle*,[2] except, perhaps, that the community of particle physicists
agrees to do so. This has not always been the case and is, in fact,
a recent development in the history of physics. The concept of a
particle, whether fundamental or not, has dissolved over time as
developments in physics have rendered the concept inherently
ambiguous.

---

[1]The grammatically incorrect title we have chosen for this short chapter communicates
the ambiguous way in which the concept of a *particle*, inherited from the classical
concept of an *atom*, is treated in contemporary particle physics. This title pays
homage to the late Sidney Drell's 1978 article, "When Is a Particle?" [*Physics Today*,
31, 6 (1978): 23–32], in which that American physicist examined the fundamentally
different considerations that arose in particle physicists' identifications of the
neutrino and quark as real particles. In this chapter, we will be thinking primarily
about the fundamental building blocks of matter: the proton, neutron, and electron.
[2]The adjective *fundamental* may be used interchangeably with *elementary*. Both
terms recover, to some extent, what has been lost of the original meaning of the Greek
root for the word *atom*.

From Atoms to Higgs Boson: Voyages in Quasi-Spacetime
Chary Rangacharyulu and Christopher Polachic
Copyright © 2019 Jenny Stanford Publishing Pte. Ltd.
ISBN 978-981-4800-24-4 (Hardcover), 978-0-429-02765-9 (eBook)
www.jennystanford.com

Nearly 3000 years ago, natural philosophers conceived of the basic material of the physical world in terms of four or five fundamental elements. The candidates were earth, air, water, fire, wind, metal, wood, and *quintessence*[3] or *aether*. Which four or five elements were accepted by the community of philosophers or larger society was closely linked to many factors, including their cultural and religious beliefs, mythologies, understanding of physical processes, and geographical context. The microscopic structure of the elements required further consideration, and two competing views were commonly held: a *plenary* theory, in which matter contains no voids and can be divided into smaller pieces without end; and an *atomic* theory, in which the elements can be broken down only to a certain finite limit, where we encounter the fundamental pieces that form physical things: *atoms*.

As discussed in the first chapter of this book, the Greek atomists are now remembered as the originators of this atomic theory, and Democritus is renowned as the chief expositor. A few characteristics of his atoms may be reviewed:

a. Atoms are indivisible[4] and stable.
b. Atoms are not all identical, but may exist with tremendous diversity.
c. Atoms of the same type, or species, are identical in all their properties.
d. Atoms may have finite sizes and shapes, and even internal properties.

This last point is worth emphasizing, in that Democritus' school of ancient particle physics did not stipulate that atoms must be point-like and structureless.

The essential characteristic of the classical atomic particle is its indivisibility, which suggests an associated *stability* as one of the properties of these building blocks. It is an unfortunate consequence of history that the word *atom* now refers, in English, to a unit of a chemical element, understood to be a composite body of smaller entities that has no essential permanence, being susceptible to ionization, radioactive decay, and nuclear fission. But neither the

---

[3]*Quintessence* was the stuff of the celestial heavens in Aristotle's cosmology. In the thinking of this Greek philosopher, the heavens were a place of incorruptible perfection and were thus made of something non-terrestrial—a *fifth element*, the literal meaning of the word.

[4]As mentioned earlier, the word "atom" means, in Greek, "uncuttable."

modern redefinition of the word, nor the antiquity of Democritus' original idea, should be allowed to deprive us of the importance of the atomic concept. With a little care, we can still use the term in its original and literal sense to denote whatever fundamental building blocks of matter our physical investigations finally uncover.

The ideas of Democritus and the other ancient atomists were, remarkably, put on firm ground thanks to the experimental science of men such as Robert Boyle, Antoine Lavoisier, and John Dalton in the seventeenth through nineteenth centuries. Dalton, the English chemist, was fully cognizant of the ancient atomic hypothesis and applied it with scientific rigor, investigating the atoms' relative weights as the principle characteristic accessible to measurement. Already in 1808, Dalton identified 20 "simple" atoms as "ultimate particles," beginning with hydrogen and ending with the heaviest known to him, mercury. These were understood to combine with other ultimate particles, both of their own and different species, to form more complex molecules.[5]

Since that time, the quest for new atomic varieties by various experimental techniques has culminated in the present-day number of 118 distinct chemical elements,[6] each with its own distinct elementary atomic unit. If *stability* were the main criterion to positively identify the discovery of a new element, the periodic table would have stopped at lead or bismuth. The most stable isotope of bismuth has a half-life of about 20 billion billion years—about twenty billion times longer than the estimated age of the universe, but nevertheless technically unstable.

7.2 The Divisible Chemical Atom

Until near the end of the nineteenth century, chemical atoms were synonymous with fundamental building blocks. This changed with the discovery of the electron, a charged particle easily ejected out of chemical atoms. This immediately disqualified Dalton's chemical atoms as the indivisible, basic units of matter satisfying Democritus'

---

[5]John Dalton, *A New System of Chemical Philosophy*, vol. 1 (London: R. Bickerstaff, 1808), 218–219.

[6]The elements above the atomic number 92 (uranium) do not occur in nature. They have been produced by artificial means at laboratories in Europe, Russia, America, and Japan and are very short lived. Of course, these discoveries do not meet the criteria of Democritus' atoms, but we may grant them status as chemical atoms.

criteria. Further, because chemical atoms are known to have zero electric charge, the discovery of the negatively charged electron implied that there should be a counterpart of positive charge. It did not take long to determine that this was the relatively massive proton, initially recognized to be the ion of the lightest chemical atom, hydrogen. Understandably, the proton and electron were quickly elevated to the status of fundamental building blocks of matter with their discovery in the early twentieth century.

Initial attempts to model the internal structure of atoms assumed that equal numbers of protons and electrons are spread over an entire atomic volume, satisfying electrical neutrality. As protons are about 1836 times heavier than electrons, the latter make a negligible contribution to the total mass of chemical atoms, leaving the protons to account for nearly all of this measurable property. This model was short lived, as the experiments of Ernest Rutherford and his colleagues, between 1909 and 1911, established that all the positive charge and almost all of the atomic mass are concentrated in the center of an atom at, essentially, a single point: the *atomic nucleus*. Later investigation would show that the nuclear volume is about one thousand-trillionth $(1/10^{15})$ the size of the overall atomic volume. The electrons were relegated to the exterior of the atom like planets orbiting a star.

At this point, there was an additional problem with the way chemical atoms were conceived: The protons, alone, could not account for the total observed mass of atomic nuclei. A heavy, electrically neutral particle was required to contribute the additional mass while keeping the overall electric charge of atoms neutral. The neutron was discovered in 1932, satisfying this demand.

Thus, protons and neutrons are now known to give mass to the nucleus of a chemical atom. Protons fix the rank of an atom in the Mendeleevian periodic table of elements, while electrons guarantee zero overall electric charge and are responsible for all chemical reactions and thus the complex biological phenomena of life. It would be reasonable to think that having identified three basic constituents that suffice for the formation of chemical atoms, we might have found the true physical entities of Democritus' philosophy. The proton, neutron, and electron certainly appear to be a complete set of particles necessary to build matter and explain the world as it is around us. However, almost immediately, the status of the proton and neutron as "fundamental particles" came under scrutiny.

7.3 Protons and Neutrons Are Particles, but Are They Fundamental?

Physicists of the twentieth century had several concerns with identifying protons and neutrons as fundamental: the finite size of the nucleons, the instability of free neutrons, and the presence of excitations similar to those seen in composite bodies such as atoms.

The first difficulty is a philosophical one: Protons and neutrons are not point objects, which is another way of saying that they appear to have definite size. Indeed, the electromagnetic properties of protons and neutrons do not conform to the theoretical predictions of point-like entities, but it should be said that even in recent years, there has been intense discussion among particle physicists about discrepancies in the size of the proton as measured by different experimental techniques.

It is not very clear when pointlikeness began to be assumed as a necessary characteristic of fundamental entities[7]: it may be that it arose through Einstein's special relativity, which assumes point-like elementary objects. Perhaps it is by inspiration of the electron, the only subatomic particle that is still universally agreed to satisfy the criteria of fundamental building block and is known to be very small in size, with a diameter of less than a billion-billionth of a meter. Whatever the case, this has come to be an important assumed property of fundamental entities in modern physics, as spatial extension has somehow become equivalent to divisibility, even though it was never a part of the Democritian atomic concept.

In some sense, infinitesimal size may just be a mathematical concept, and the acceptance of real point particles may thus be a quasirealist interpretation. Physicists are known to start off with mathematical simplifications and interpret physicality from there. Even at the early stages of development for gravitation and electricity theories, it was known through the works of Carl Gauss

---

[7]We came across this in the English edition of the celebrated book, *The Classical Theory of Fields,* by L. D. Landau and E. M. Lifshitz (Pergamon Press: 1971).

that uniform mass or electrical charge distributions within a sphere can be approximated as a point mass or point charge located at the center of the sphere. For all practical purposes, the external world outside the sphere is insensitive to whether the object is a point or of finite size.

This Gaussian conceptualization is an *effective* one: a simplification for mathematical purposes that works so long as the observer or the interaction partner is physically outside the volume occupied by the actual charge or mass distribution. If, however, we take the simplification too seriously and then ask about the mass or charge density of the apparent point-like body, we will wind up with a physically meaningless infinite value as the answer. Results like that should guard us against the error of quasirealist interpretations of the mass or charge distribution, i.e., thinking that the effective mathematical entity had claims upon reality.

Remarkably, these kinds of infinity problems have been plaguing modern particle physics from its beginnings. Quantum mechanical and quantum field theories have long suffered from them. They have been addressed through mathematical *renormalization* procedures, involving the unsettling subtraction of infinities in order to arrive at finite results. If anything, this continuing problem might suggest the necessity of finite size for the ultimate building blocks of nature.

Besides the concern over the finite size of nucleons,[8] it is well known that neutrons, outside the nucleus, are not stable, but have a half-life of about 10 min. The weight of this objection against the neutron being a fundamental particle is reduced by the inconsistency of its application (as we discuss later, *stability* is not understood to be a necessary property for a particle in modern physics) and the reality that the products of neutron decay are themselves stable entities. Certainly, the instability critique does not apply to protons, which, if not fundamental entities, have the awkward physical reality of being composite bodies that never break into their pieces!

The third concern with treating the nucleons as fundamental, like the electron, is the presence of their energy excitations. As a general understanding, an entity can be excited to higher internal energies only when it has internal structure to facilitate the change

---

[8]Nucleon is the collective name for neutrons and protons, referring to either one.

in internal energy.[9] Again it should be recalled that this is only an argument against indivisibility, and not fundamentality unless it is assumed that fundamental particles must have no internal structure or finite size. But those restrictions would require justification.

As things stand, then, we would suggest that protons and neutrons may yet qualify as irreducible building blocks of ordinary matter, despite their finite size and internal properties. Indeed, we will reinforce this suggestion in the next chapter by questioning the ontological status of quarks, arguing that this status may well be entirely dependent on quasirealist interpretation of mathematical formalism. If that is the case, then protons and neutrons might be true atoms in the Democritian sense.

7.4 The Electron Is Fundamental, but Is It Still a Particle?

The electron was the first particle discovered. It was observed in experiments performed by J. J. Thomson[10] in the late nineteenth century[11] and was soon understood as an essential constituent of atoms, along with protons and neutrons. In contrast to the nucleons, electrons have thus far retained the esteemed distinction of being fundamental: Today they are accepted by everyone as truly elementary bodies, not composed of any smaller components.

Electrons are the lightest charged particle. An isolated or "free" electron has a mass of 0.511 MeV/c^2 (9.11×10^{-31} kg) and one negative unit of the fundamental electric charge (-1.602×10^{-19} coulomb). These properties are well established through straightforward experiments and provide the necessary inputs to determine the kinematical behavior of electrons within Newtonian theory.

---

[9]Nuclei, upon excitation, may disintegrate to lighter counterparts wherein neutrons and protons may be converted from one to another, but the total number of nucleons must remain the same. For example, a uranium atom may fission to lighter elements and the sum of the protons and neutrons in those resultant atoms will be the same as in the original uranium atom. Similarly, a neutron decay will yield a proton such that the total number of nucleons remains constant.

[10]Joseph John Thomson (1856–1940) was an English physicist. He was awarded the Physics Nobel Prize in 1906 for his study of electrical conduction in gases, and devised a "plum pudding model" of the atom that was subsequently replaced by Rutherford's planetary model.

[11]J. J. Thomson, "Cathode rays," *Philosophical Magazine*, **44** (1897): 293–316.

7.5 The Electron of Wave Mechanics

The advent of quantum theory in the early twentieth century radically changed the concept of a free particle, and the electron was the first entity to suffer the ontological consequences. As is well known, Niels Bohr introduced an early quantum model of the hydrogen atom in which the bound electron can only occupy discrete orbital states around the nucleus, transitioning between these states in quantum jumps associated with specified, allowed changes to its angular momentum state, and the absorption and emission of discrete frequencies of light. Within these stable orbits, the electron would defy the expectations of classical electrodynamics by never radiating away energy due to its constant change in direction of travel. On the surface, the assumptions of Bohr's model appear artificial, designed as *ad hoc* postulates that nonetheless provide a sufficient physical model to explain some real experimental data relating to absorption and emission of light by atoms. But the jarring characteristics ascribed to electrons in the atom require explanation.[12]

A possible way forward came from the doctoral thesis of Louis de Broglie. Inspired by Albert Einstein's work on light quanta, de Broglie took a conceptual leap by suggesting to his supervisory committee that if light can have definite properties of particles, then in all fairness particles of matter should be allowed to have definite wave properties as well. It was certainly a revolutionary and bold idea but seemed like fantasy to his reluctant doctoral committee. Paul Langevin, de Broglie's supervisor, sent a copy of the thesis to

---

[12]Robert Millikan provided an early justification for the stability of the electron orbits that now seems prescient of the matter wave concept that would soon follow. In his 1917 book, *The Electron* [1st edn. (Chicago: University of Chicago, 1917; repr., Chicago: University of Chicago, 1966)], Millikan wrote approvingly, "N. Bohr, a young mathematical physicist of Copenhagen, has recently devised an atomic model which has had some very remarkable success" (207). He noted that "its chief difficulty arises from the apparent contradiction involved in a non-radiating electronic orbit—a contradiction which would disappear, however, if the negative electron itself, when inside the atom, were a ring of some sort, capable of expanding to various radii, and capable, only when freed from the atom, of assuming essentially the properties of a point charge" (216). Millikan's proposal, in contrast to the view ultimately adopted in the quantum theory, seems fully situated within a classical conviction that the components of matter have definite physical structure of one kind or another, localizable in space and time.

Einstein for review, and the latter assured the former that it was a work of great insight.

Putting aside how unusual it is, in itself, this matter–wave thesis provided a useful way of picturing the unusual properties of atomic electrons. If electrons are not thought of as particles orbiting a nucleus, but somehow as a wave that must have a wavelength measured around the nucleus that is mathematically consistent with its energy and orbital radius, then quantized orbits and discrete frequencies of absorbed and emitted radiation fall out of the calculations quite readily. Within a few years, experiments seemed to indicate that a beam of electrons incident on a metal surface do, indeed, exhibit characteristic wave behavior like diffraction.

De Broglie's hypothesis provided the inspiration for the next great leap in the development of quantum theory: *wave mechanics*. Erwin Schrödinger took notice of Einstein's support for the matter–wave concept and devised his famous equation describing the evolution of matter waves in time. This *Schrödinger equation* is designed to calculate the way the energy of a particle, like an electron, evolves, taking into account both the particle's motion and the effects from external forces that enter the equation through their associated potentials. The particle, itself, is represented by a mathematical function called the *wave function*, which describes all its relevant physical properties in a compact way.

In keeping with its name, the wave function has a mathematical structure that explicitly assumes the wave properties of de Broglie's hypothesis. It is here that we have an essential deviation from Newtonian concepts of particle behavior, as foreshadowed in the previous chapter. In the classical theory of mechanics, a body in motion will experience undulatory behavior only when imposed upon by a time-varying external force. In other words, wave behavior always arises due to some external periodic oscillation to which an elastic medium responds. There must be definite external forces present that push or pull with some frequency, and only then is it sensible to speak of a particle of matter participating in wave-like phenomena. Most often, in fact, wave behavior is a bulk phenomenon, the collective response of interacting molecules that mechanically transmit the effects of an external, time-varying force through a medium.

This essential conceptualization of wave motion in matter is dispensed with in Schrödinger's wave mechanics, where the wave function and Schrödinger's equation introduce something extraordinarily new: oscillatory behavior as an inherent property of a particle, divorced from a causal force! What appears as a fundamental violation of Newton's first, inertial, law of motion becomes a unique feature of the way matter is described in the quantum theory. In fact, in the Schrödinger equation, the electron oscillates whether or not there is a nonzero potential term present. In the context of the Bohr atom, which works well with de Broglie's matter wave hypothesis, one may feel justified assuming the strong, central electrostatic force of attraction between the electron and the nucleus somehow provides an oscillatory aspect to the particle's behavior. In that case, the wave function would be an *effective* description of this reality: The oscillation is mathematically transferred to the functional representation of the particle. However, in Schrödinger's picture, this is not understood to be the case, for his equation is not limited to bound systems like atoms. It applies generally, providing time-evolution information even for a free electron whose environment contains no forces.

This leads to a problem of physical interpretation, taken up by the early interpreters of quantum theory. If the wave function describes the electron *as it really is*, what is waving? What can be oscillating, in time and across space, if no forces act to make it so? Whatever answer we arrive at will take us far away from Newtonian dynamics. Physicist Max Born provided the now widely accepted answer, in the form of a further abstraction of the electron's nature: the wave function is related to the probability of the electron being in a particular state (including spatial location) when measured at some time.[13]

Physicists and philosophers of science have understood the meaning of Born's rule in different ways. On the one hand, the language of probability suggests that wave mechanics can only tell us something about the statistics resulting from repeated measurements on identical individual systems: The reality of the electron remains buried under mathematical averages. On the other hand, the wave function and its connection to probability might

---

[13]Stated more precisely, the square of the magnitude of the wave function of an electron provides the probability of the particle's state.

be interpreted to mean that each electron, on its own, exists in a condition of probabilistic fuzziness, simultaneously occupying all the possible states described by the wave with some of its realities more dominant than others, according to the relevant statistical weights.

It is a remarkable fact of scientific history that the latter interpretation of the meaning of wave mechanics has now won the field in terms of how physicists are taught to understand quantum theory. Electrons, and indeed all matter, are now understood to exist in a kind of phantom reality, a mathematical superposition of states each as real as the other, but all simultaneously valid. Understandably, a commitment to this view endows quantum theory with many consequences that defy common sense, especially the difficult and famous phenomenon of *quantum entanglement* and the breathlessly referenced "spooky action at-a-distance" that occurs between mathematically entangled systems. The conceptual impossibility of making sense of quantum mechanics is the modern physicist's badge of honor in public discourse.

7.6   Niels Bohr's Instrumentalist View

The confusion of these quantum ideas is in tension with the simple realism behind the words that are still employed in the theory. In particular, the idea of a *particle* and its *position* or *momentum* in space has experienced a fundamental alteration as physicists have attempted to uncritically embrace quantum wave mechanics as a proper description of *how physical reality really is*. Since the beginning of quantum theory, it has been clear that scientific realism was undergoing a dramatic challenge. For the more thoughtful physicists of the day, like Niels Bohr, this resulted in an unapologetic disposal of realist assumptions and the adoption of an instrumentalist philosophy of physics in the form of Bohr's now well-known, but universally confusing, *Copenhagen interpretation* of the theory.

The Copenhagen interpretation requires only that the mathematical statements of quantum physics somehow support one or another outcome of a measurement, without requiring an essential correspondence to a unique, real property of the system. In some experiments, electrons really seem to be spatially localized particles;

in others, they seem very much to be extended, indefinite probability waves. To ask, "which is the true nature of the electron?" is to ask the wrong question, according to the Copenhagen interpretation. It is, in many ways, a resignation to our ignorance—an admission that if we embrace the wave function as the way electrons really are, then our classical, realist intuitions will simply have to be set aside and there is no use in protesting.

For the scientific realist, Bohr's approach is disappointing but respectable—a sensibility that he may have felt: "In spite of the fruitfulness of quantum mechanics within such a wide domain of experience," he admitted, "the renunciation of accustomed demands on physical explanation has caused many physicists and philosophers to doubt that we are here dealing with an exhaustive description of atomic phenomena."[14] Whatever else, though, Bohr's view was an honest one, for it is certainly within the realm of possibility that the essential structure of nature is beyond human logic at some fundamental level.

Bohr's stance was not quasirealism, for he explicitly recognized that the mathematical abstraction of this theory rendered it realistically suspect:

> In order to obtain a consistent account of atomic phenomena...as in the formulation of relativity theory, adequate tools were found in highly developed mathematical abstractions. The quantities which in classical physics are used to describe the state of a system are replaced in quantum-mechanical formalism by symbolic operators whose commutability is limited by rules containing the quantum.[15]

Of particular concern to him was the use of the imaginary $i = \sqrt{-1}$ in both quantum theory and relativity, which made these theories alike in being non-literal representations of the system of nature:

> Notwithstanding all differences between the physical problems which have given rise to the development of relativity theory and quantum theory, respectively, a comparison of purely logical aspects of relativistic and complementary argumentation reveals striking

---

[14]Niels Bohr, *Atomic Physics and Human Knowledge* (New York: Science Editions, Inc., 1961), 88.
[15]Bohr, 87.

similarities as regards the renunciation of the absolute significance of conventional physical attributes of objects.... The astounding simplicity of the generalization of classical physical theories, which are obtained by the use of multidimensional geometry and non-commutative algebra, respectively, rests in both cases essentially on the introduction of the conventional symbol $\sqrt{-1}$.[16]

According to philosopher Jan Faye, Niels Bohr is best understood as "an antirealist or an instrumentalist"[17] in his understanding of the extent to which non-classical physics provides insight into the real entities of nature.

The historical choice of the physics community was to eschew Bohr's instrumentalist view of quantum theory and rather cling—awkwardly—to the impossible task of uniting the Born interpretation with primordial realist convictions. The awkwardness comes from insisting, to ourselves, our students, and the listening public, that every aspect and result of the theory reflects *how things really are*. This is an exemplar of quasirealism, and the consequences for the electron are unfortunate: An entity that appeared in early experiments with physical definiteness was transformed into a phantom hybrid through a fundamentalist reading of the sacred text of quantum theory.

7.7   Electrons in Quantum Electrodynamics

Besides the mass and electric charge of a particle, the next important classical attribute is the *magnetic dipole moment*, a measure of an electron's response to an external magnetic field. Particles like electrons that have nonzero intrinsic spin are nature's fundamental magnetic entities, and their behavior in a magnetic environment will be governed by this property.

Nuclear and particle physicists have developed elaborate mathematical schemes to calculate electric and magnetic properties of particles and nuclei, both for those of finite sizes and of various shapes. In all these theoretical schemes, we take a point object as a

---

[16]Bohr, 64–65.

[17]Jan Faye, "Copenhagen Interpretation of Quantum Mechanics," ed. Edward N. Zalta, *The Stanford Encyclopedia of Philosophy (Fall 2014 Edition)*, https://plato.stanford. edu/archives/fall2014/entries/qm-copenhagen/.

reference, calculate its properties for specific mass, charge, and spin parameters, and estimate how much the measured values deviate from these. The only way to know if the theory is reasonable is to compare with an experiment measuring interactions with external forces and fields. We can let the system be influenced by an external electromagnetic field, or, even worse, we can make the system undergo transitions between its characteristic structural states. Then we draw quantitative conclusions from these measurements within the framework of our favorite theories and models.

Quantum electrodynamics (QED), an advanced version of quantum theory, provides the framework for these calculations. QED incorporates aspects of Einstein's special theory of relativity to describe the interactions of particles like the electron with electric and magnetic fields and is hailed as the most rigorous and precise theory of these phenomena. These interactions are modeled by a series of *exchanges* between the particle and bosons, which transmit the field's affects. The exact details of the interaction mechanism can become quite complicated, and the associated mathematics increasingly time-consuming and difficult, but the extra effort occasionally yields an output of the theory that fits the experimental data with increasing precision. In the last 60 years or so, a great deal of theoretical and experimental effort has been invested, by many people, toward this end. The motivation has been twofold: first, to establish QED as a precision theory that can be applied to arbitrarily large- and small-scale phenomena; second, to set the limits at which QED breaks down, always aware that this eventuality may shed light on "new physics."[18]

Applying QED to studying the electron's magnetic moment has a long history of over 70 years, involving a few highly eminent physicists who dedicated their careers as experimentalists and theoreticians to the task of taming this parameter. From the perspective of theorists, the magnetic moment depends on three parameters: the particle's mass, charge and a parameter we call "*g*" (which is a function of intrinsic spin), as well as Planck's constant and the speed of light.

---

[18]The phrase "new physics" has become a buzz word in the last few decades, without any necessary specification of its meaning. The basic idea is to find some discrepancy between the predictions of the Standard Model of particle physics and an experimental result, thus challenging the theoreticians to develop those models.

Both theory and experiment agree on the value of the electron mass and charge, so it is g that comes in for comparison between the two. How might this parameter be measured in an experiment? Clearly, the physicist ought to identify a relevant observable that will ultimately provide the appropriate number. The energy gain or loss of an electron in an external magnetic field provides such an observable, as it is sensitive to the precise value of g and is accessible to both experimental measurement and theoretical calculation.

The most rudimentary application of QED to a specification of g for the electron magnetic dipole moment was first worked out by Paul Dirac.[19] This assumed an infinitesimal size for the electron and gave a value $g = 2$, which agrees with experiment to about one part in a thousand: The earliest *experimental* measurement from 1948 resorted to atomic-level splittings in the magnetic field[20] and yielded a value of 2.00232. This correspondence is an impressive result for QED, but it is still worthwhile to explain the small discrepancy.

Dirac's prediction of the magnetic moment simply assumed that an electron absorbs energy from an external magnetic field through interaction with a photon. The discrepancy raised a concern, not with the plausibility of electrons as fundamental particles, but about the behavior of electrons in the presence of external fields and the way they propagate in spacetime from point A to point B. If electrons interact with these fields through exchange of photons, perhaps more than one photon is involved. In QED, one asserts that these are not real photons, in the sense that they cannot be observed in the laboratory. They are referred to as *virtual photons*, and the more that are involved in the interaction process, the less likely it is to occur.

Julian Schwinger[21] attempted to improve on Dirac's single-photon-exchange calculation by invoking what sounds like a strange

---

[19]Paul Adrian Maurice Dirac (1902–1984) was a British–American physicist. He was a giant of mathematical physics in the twentieth century and the founder of quantum electrodynamics. Dirac is famous for using his theory to predict the existence of antimatter in 1928.

[20]Even before the advent of quantum mechanics, it was known that the atomic and molecular spectral lines split into more components when the samples are in external magnetic fields. This phenomenon was called the Zeeman effect, and it contributed to the development of the intrinsic spin concept.

[21]Julian Schwinger (1918–1994), an American physicist, well known for contributions to QED.

complication to the interaction: The electron absorbs the photon but then simultaneously emits a virtual photon and in the process becomes a virtual (non-physical) electron. This emitted photon is then reabsorbed by the virtual electron to render it physical once again. These intermediate, virtual particle states are not required to satisfy the foundational energy conservation rule of Newtonian and relativistic physics, thanks to the development of Heisenberg's uncertainty relation. Schwinger's mechanism, complicated as it is, provides increased agreement between the mathematical output of QED and experiment: better than one part in one hundred thousand! It brought the theoretical and experimental values of the day into good agreement.

But there is no need to stop at this point with QED, for there is a set of rules that a theoretician can follow to expand their calculations and include ever more complicated virtual states in that basic interaction of the electron with the field. These can be illustrated using *Feynman diagrams*, convenient sketches that illustrate the interaction of particles over time. Figure 7.1 shows a few diagrams that are relevant to the processes being considered: The electron interacting with the field is indicated by a straight line, and the wavy lines are virtual photons. The simple Dirac process at the top, with Schwinger's advancement underneath. Moving down the figure, additional steps may be added of the electron emitting and reabsorbing multiple virtual photons. Alternatively, in the fourth diagram, the electron may emit a virtual photon, which then transforms momentarily into a particle–antiparticle virtual pair (the circle) of zero total charge, which loop around and annihilate one another again to become a virtual photon, which is reabsorbed into the original electron, restoring it to physicality. In this loop scenario, a remarkable thing happens in our electromagnetic interaction: Weak and strong nuclear forces make an appearance in the mechanism!

The possibilities for more and more complications are literally endless, and the quasirealist understanding of these possible events is that every time an electron interacts with an electromagnetic field, every one of the (infinite) possible interactions actually takes place simultaneously, in the being of the electron. Only by adding up the mathematical contribution of each possibility in QED will we know whether the theory can truly recover a value for the magnetic moment that perfectly agrees with experiment. But it is a game of diminishing returns, for there exists a rule of thumb that the more complicated the set of interactions, and thus the more difficult the calculations, the less important are their overall contribution to the final number.

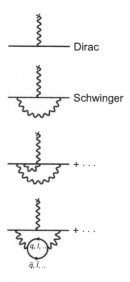

Figure 7.1 Illustration of a few Feynman diagrams of current interest to calculate the electron magnetic dipole moment. The more lines and loops mean a more involved calculation.

This results in a race between theory and experiment. Experimentalists will try to take measurements to the best possible precision and continually improve upon the last result to get even better numbers. Theoreticians take up the challenge of including higher and higher order interaction events to deduce these numbers to better precision. If the theory, as accepted, does not ultimately overlap with experiment, then it is interpreted as a result pointing to new physics. The goal of this exercise seems to be to improve upon experimental precision until an inevitable stage is reached where theory fails. If it does not fail after a million complications are introduced into the life of an electron, try to add a billion more abstract virtual interactions, confident that eventually the theory will break down. This general philosophy of testing a theory to death is not unique to the problem of calculating dipole moments.

Currently, the electron magnetic dipole moment calculations are carried out to about the eighth order of corrections to Dirac's original value. While the first-order correction could be done by a competent undergraduate student, these refinements now require several decades of dedication from large groups of intellectuals with access to frontier technologies for the experimental work, and excellent analytical techniques as well as powerful computational resources for the theory. At the end, is there a simple mental picture of what the electron's magnetic moment property is? Can it be explained even to their colleagues, let alone to professionals in other scientific communities, or to the general public? Does QED provide us with an increasing sense of what is real about the electron? Perhaps, instead, its increasing complexity and demand for mathematical abstraction simply points to the fact that QED and the underlying foundation of quantum theory are useful instruments, up to a point, and nothing more.

It is worth considering again that the deviation of the electron magnetic moment from rudimentary theory was never understood in what might have been a rather obvious way: as an indication that an electron may not be a fundamental point particle. The neutron magnetic moment is expected, theoretically, to be zero since it is electrically neutral. We also expect the g factor for the proton to be 2, the same as with electrons. Experiments, however, found the neutron g to be −3.82 and that of proton to be 5.82. When the proton and neutron moments[22] were found to differ in this way from rudimentary elementary particle estimates, the deviation was attributed to the compositeness of these entities, not to higher-order corrections to the interaction process that are as-yet uncalculated. It is not clear that electrons and nucleons have been treated in a consistent way in interpreting the physical meaning of their magnetic moment anomalies.

7.8   Electrons in Bulk Matter

There is a useful contrast to the quasirealist misuse of the Born interpretation, found in the way electrons are treated in the study

---

[22]It is worth noting that the proton and neutron magnetic moments are smaller in magnitude by a factor of about 1/1000th to that of electron.

of bulk matter. The study of matter in its solid state involves an exploration of a material environment of great complexity, nonetheless organized according to some remarkable underlying rules of symmetry. Together, the symmetry and complexity of many classes of bulk material allow for the use of simplifying mathematical assumptions that provide extremely useful *effective* models of electron behavior in these environments.

The study of bulk matter is motivated by obvious technological benefits, and early solid-state physicists made concerted efforts to describe the thermo-electrical properties of idealized real material media. This was greatly simplified by the microscopic structure of relevant systems: atoms of many important materials have a periodic ordering and spacing with respect to one another, and this is readily captured by the same kind of mathematics that appears in periodic descriptions of wave phenomena. The wave function description of the electron was usefully adopted into and melded with a mathematical description of the atoms that make up the scaffolding of bulk materials, otherwise known as the *crystal lattice* of solid matter. This melding of periodicities results in a remarkable new entity appearing in the mathematics of solid-state physics: *quasiparticles*.

Quasiparticles are purely mathematical entities, but they are useful because they appear in the formulas like real particles. They arise from the mathematics used to describe the collective behavior of all the electrons that inhabit a crystalline solid (such as a conducting metal) and interact with the many positive nuclei laid out around them. Through a series of mathematical transformations, the equations that collectively describe all the real electrons can be re-expressed in terms that appear to deal only with individual entities. What is most remarkable is that these quasiparticles have mathematical properties very much like individual electrons— mass, momentum, and electric charge—and behave as if they are free particles rather than electrons living in a periodic crystal lattice of positive charge. The ionic centers of a solid are thus absorbed into the mathematics of the quasiparticle description and averaged away leaving a collection of electron-like entities that are similar enough to the real things that solid-state physicists easily refer to them simply as the "electrons" in the system.

The quasiparticle model has been extraordinarily useful but produces some results that would be surprising, indeed, if we forgot that these are not real electrons. For example, in some solid-state models, quasiparticle masses can be many times that of the free electron mass and can even take on negative values. The mass of a quasiparticle is *usually* referred to as an *effective mass* to remind us that it is not the real mass of a real electron. According to Newton's second law of motion, a force (F) acting on a free electron of mass m will result in an acceleration (a) of that body according to the following relationship:

$$F = ma \tag{7.1}$$

This mass can be measured by applying an electrostatic force of the form

$$F = eE \tag{7.2}$$

where e is the charge of the electron and E is the electric field associated with the electrostatic force. Thus, the acceleration of the electron yields the mass by equating these forces and solving for m:

$$m = \frac{eE}{a} \tag{7.3}$$

An electron in a solid material, however, is not free: It is subject to a complex background of forces from all the ions and other electrons in its environment. It is impossible to capture the details of these forces in a precise, physical mathematical expression, so one resorts to a quasiparticle treatment. The additional background forces are summed up as a variable F_b, which are added into the force on the electron in Eqs. (7.1) and (7.2) in the following way:

$$F + F_b = eE + F_b = ma \tag{7.4}$$

It now seems that the electron's response to an applied electric field will be

$$eE = ma - F_b\left(\frac{a}{a}\right) = m^*a \tag{7.5}$$

where the effective mass is a parameter that incorporates not only the actual electron mass, but also the background effect of other forces:

$$m^* = m - \frac{F_b}{a} \tag{7.6}$$

This quasiparticle mass may take all kinds of unexpected values, including negative ones. In a similar manner, there are quasiparticles that acquire fractional electric charge in the context of a phenomenon called the quantum Hall effect. Properly interpreted, these attributes help physicists to understand experimental data, but they are not understood as the mass or charge of ontologically real particles.

The solid-state quasiparticle model has been expanded beyond effective electrons and is even used to represent the vibration of atomic nuclei. Under a suitable mathematical transformation, the vibrational modes of nuclei can be represented as particles called *phonons*, but it is well-understood that this, too, is just a sleight of hand: No self-respecting condensed matter physicist incorporates the idea of physical phonons into their realist understanding of matter.

In fact, it is to the credit of these scientists that, despite several decades of immersion in the quasiparticle framework, they do not regularly claim to have discovered new forms of fundamental matter each time they solve one of their models and find a unique quasimass, quasicharge, or quasimomentum. Indeed, such claims are occasionally made in the press by uninitiated journalists and web editors. A recent example of this was *Newsweek's* headline, "Particle Physics: 'Mind-Bending' Negative Mass Device Reveals New Way to Create Lasers," followed by the opening sentence, "Physicists have designed the first device to create particles with charged negative mass."[23] It is gratifying to see the relevant research article in the journal *Nature Physics* put the blame for the phenomenon solely at the feet of "many-body effects" that result in "an effective level attraction between the exciton-polariton and trion-polariton accounting for the experimentally observed inverted trion-polariton dispersion"[24]—this is jargon for quasiparticles that arise from the bulk properties of the material. The word "effective" is key. Every graduate student in condensed matter physics knows better than to confuse these with real particles, even though they have the

---

[23]Kastalia Medrano, "Particle Physics: 'Mind-Bending' Negative Mass Device Reveals New Way to Create Lasers," Newsweek, January 12, 2018, http://www.newsweek.com/particle-physics-mind-bending-negative-mass-device-lasers-778495.

[24]S. Dhara, C. Chakraborty, K. M. Goodfellow, L. Qiu, T. A. O'Loughlin, G. W. Wicks, S. Bhattacharjee, and A. N. Vamivakas, "Anomalous Dispersion of Microcavity Trion-Polaritons," *Nature Physics*, **14** (2017): 130.

mathematical structure of particles and follow seamlessly from the rules of quantum mechanics applied to the solid state.

Despite the challenges of public communication, there is a way to employ mathematically abstract physics for its usefulness, but not be led away into quasirealist interpretations. Condensed matter physics provides an example that particle physics might emulate.

7.9   In Summary: We May Still Have Atoms

It is long since physics lost sight of the classical atomic theory in which the basic pieces of matter are indivisible bodies. Despite the unfortunate retention of the atomic label, the chemical atoms made of nuclei and electrons certainly do not qualify as these building blocks. If anything, indivisibility has been replaced by the dubious quality of infinitesimal size, disqualifying protons and neutrons as well. In this light, only the electron retains a claim to the ontological status of atomicity among the basic components of the matter of everyday life.

But the advent of quasirealist interpretations of quantum theory has taken even the otherwise well-behaved and simply understood electron and turned it into an ontological mess. When modern physicists speak of "electrons," they are no longer talking about the entity observed by Thomson, but a quasiparticle, although his concrete demonstration of their physical reality still lingers in our peripheral vision. As a result, we have learned to think of these quasiparticles as exact mathematical representations of that real subatomic entity, despite the clear divergence of properties. In this chapter, we have tried to show that the similarity is only due to historical continuity and not essential quality.

We need not, however, abandon the idea that nature is composed of real Democritian atoms, including real electrons, but they are not the quasiparticle electrons of our textbooks. Neither are they the *quarks* and *gauge bosons* of the modern particle physics model. Even though quarks and gauge bosons are currently understood to comprise the fundamental building blocks of nature, in the next chapter we will argue that this, too, is probably a quasirealist misinterpretation. The electron, proton, and neutron may be the only physical entities reliably situated at the bottom of physical reality, once all the quasirealist misdirection is cleared away.

Chapter 8

Elementary Quanta

We especially need imagination in science. It is not all mathematics, nor all logic, but it is somewhat beauty and poetry.

We must at least question it; we cannot accept anything as granted, beyond the first mathematical formulae. Question everything else.

—Maria Mitchell
Maria Mitchell: Life, Letters, and Journals (1896)

8.1 Fermions, Bosons, Quarks, and Leptons

In this chapter, we will provide an overview of what modern physicists think of as the *elementary quanta*, the ultimate components of the physical world. It will be well to keep in mind our last chapter in which we focused on the electron as a potential candidate for a real, fundamental building block of matter, but also indicated how the concept of the electron has been subverted by quasirealism. We will now consider the other candidates for elementarity, examining the state of conceptual clarity that currently defines their properties and the contribution these quanta make to the dynamics of physics. There is some ambiguity as to which elementary quanta can be colloquially considered "building blocks" of *matter* or of *energy* or of some hybrid between the two. We will treat all cases in this chapter.

From Atoms to Higgs Boson: Voyages in Quasi-Spacetime
Chary Rangacharyulu and Christopher Polachic
Copyright © 2019 Jenny Stanford Publishing Pte. Ltd.
ISBN 978-981-4800-24-4 (Hardcover), 978-0-429-02765-9 (eBook)
www.jennystanford.com

Elementary quanta can be classified into two[1] distinct groups called *fermions* and *bosons*,[2] based on the symmetry properties they exhibit. These groups can be further subdivided into four families called *leptons*, *quarks*, *gauge bosons*, and the *Higgs boson*. Fermions and bosons are defined and distinguished on the basis of a property called the *intrinsic angular momentum*, more conveniently known as "spin."

Angular momentum is a very familiar property of classical objects, relating the rotational or orbital motion of a body with respect to the position in space of another. It plays a prominent role in Kepler's laws of planetary motion in our solar system, especially as they are deduced from Newton's gravitational law. Angular momentum is a contingent property of objects, always measured *relative to* some other body, and is never considered an *intrinsic* property of a classical system.

Historically, physicists borrowed the framework of this successful planetary model to describe atomic structures, anticipating a replication of the same rules in nature applied at different scales. After all, Coulomb's law of electrostatic force has the same inverse square rule for separation distance between two bodies as found in Newton's law of gravity. In a planetary model of the atom, atomic electrons behave like planets moving about a central nucleus in the place of the sun, and thus angular momentum is a useful concept in describing the relationship of the orbiting electrons to this center of attraction.

---

[1]In condensed matter physics, another family of quanta has been identified as *anyons*, which are distinct from fermions and bosons, discussed here. Condensed matter physicists understand anyons to be quasi-entities. As explained in the previous chapter, quasi-entities are mathematical constructs used to describe collective effects of an ensemble of real entities. For this reason, they are not really a category of elementary quanta and would not be discussed any further here.

[2]Fermions were named after Enrico Fermi (1901–1954), an Italian physicist who migrated to the United States. He was involved in both the theoretical and experimental sides of physics and led the group who demonstrated the first nuclear reactor. This work led to the development of atomic weapons as well as nuclear power plants and research reactors. Bosons were named after Satyendra Nath Bose (1894–1974), an Indian physicist who laid the foundation of what today is called Bose–Einstein statistics, which provides a very important contribution to understanding the rules governing so-called elementary particles.

Niels Bohr's early ideas about quantum theory began with a planetary model description of the hydrogen atom. Physicists of the nineteenth century had struggled to explain the mechanism behind the emission of discrete colored spectral lines from energized samples of elemental gas, and Bohr's interpretation of the phenomenon relied on the concept of electron-planets circulating about the nucleus in discrete, stationary, closed orbits.[3]

Bohr's model of the atom incorporates an important feature: The total energy of an electron—the sum of its potential and kinetic energies—varies in a discrete manner, corresponding to its particular orbit. Unlike real planets orbiting within our solar system, electrons cannot just orbit anywhere, but only at very specific orbital distances and energies. When electrons move between orbits in this model, they either absorb or emit energy in the form of photons that have exactly the right energy as that of the electron's atomic transition. In keeping with work done by Max Planck and Albert Einstein before Bohr developed his model, the energy of a photon should be proportional to the frequency of the radiation emitted or absorbed, multiplied by a constant called the Planck constant, denoted throughout physics in one of two mutually proportional forms[4]: by the letter h or the symbol $\hbar$.

The consequence is that electrons in these atomic orbits should carry an angular momentum proportional to the Planck constant. Quantitatively, the angular momentum for different orbits will be an integral multiple of $\hbar$, that is, $n\hbar$ with $n = 0, 1, 2, 3$, etc. In this we have the beginnings of the concept of angular momentum *quantization* from which quantum theory derives its name: The energy of electrons in the Bohr model of the hydrogen atom is *quantized*, which simply means they have discrete values differing by indivisible numerical steps. It is useful to keep in mind that angular momentum quantization is, therefore, a consequence of the assumption of bound orbits and Planck's photon energy concept. It is not a property of the electron, itself, but a contingent parameter defined with respect to

---

[3]Bohr's orbital picture was soon discarded as unphysical by Werner Heisenberg and others. However, a century later, students in modern physics and chemistry courses still encounter this orbital picture in their introductory textbooks.

[4]The Planck constant, $h = 6.63 \times 10^{-34}$ J·s, is related to the "reduced" Planck constant, $\hbar$, as $\hbar = h/2\pi$.

the electron's place within the atomic or molecular system to which it is bound.

The concept of angular momentum in this early, developing quantum theory took an interesting turn in the 1920s when experimental anomalies were observed in emission line spectra. Since at least the late nineteenth century, it was known that sodium atoms in an external magnetic field emit a characteristic doublet of bright spectral lines. This so-called *Zeeman effect* seemed to require the *ad hoc* introduction of a new physical property for atoms. A new kind of angular momentum solved the mystery.

This new property was defined as essential to the physical nature of the particle in the form of an *intrinsic* angular momentum, often called *intrinsic spin*. The spin of a quantum particle is a property that is, for mathematical purposes, identical to classical angular momentum and has the same physical consequences for interacting systems in terms of conservation rules. The difference is that spin angular momentum is simply *specified* as an essential characteristic of a type of particle, independent of environment. A particle's spin value has no reference to other interacting partners, and it does not correspond to the local rotation of a spatially extended body or a moving body's relationship to any reference point in space. Any attempt to describe this intrinsic property by analogy to classical ideas like spinning tops proves futile and misleads the student of physics: There is no classical counterpart to spin.

Not surprisingly, it took nearly 30 years for physicists to incorporate spin into their theories after the first anomalies were observed. In the late 1920s, after several years of theoretical formulations and experimental investigation, including measurements of ion deflection in magnetic fields, intrinsic spin was enshrined in physics as a mature concept. It solved the riddle of spectral line doublets and other multiplets and also described many other physical observables in a consistent manner. The concept of intrinsic spin has led to many theoretical arguments, which have stood the test of time.

Every quantum body is assigned a numerical value of its spin angular momentum proportional to $\hbar$: either an integral multiple $n\hbar$ or a half-integral value $(n + 1/2)\hbar$, where $n = 0, 1, 2, 3$, etc. as before. Half-integer spin particles, which include electrons and protons, are the fermions. The bosons are those with integer spin

(0, 1, 2, and up), including photons and mesons. Composite systems made up of a collection of particles can be likewise classified by adding up the total spin: An odd number of fermions together make another fermion because an odd number of halves is also a half-integral value; an even number of fermions or any number of bosons form a new bosonic system. This scheme of identification is a consequence of angular momentum algebra, a simple set of rules of addition of vectors and their projections onto mathematical coordinate axes.

Fermions obey rules referred to as *Fermi–Dirac statistics,* for which the *Pauli exclusion principle* is a fundamental premise. This principle insists that no two identical fermions can occupy the same physical state. By "state" we refer to the complete set of properties of the fermion, such as its position in space, angular momentum, or other more exotic aspects of its identity within quantum theory such as its isospin, QCD color, etc. Each property of a particle's state is evaluated by a *quantum number*. Another way of looking at the Pauli exclusion principle is to see that no two identical fermions can share all the same quantum numbers. This principle has profound implications for the structures of nuclei, atoms, molecules, and more complex entities.

There is a very fundamental law of physics that any physical system, left to itself, will settle into the lowest possible energy state available to it. The exclusion principle prevents two fermions from occupying the same energy state, so once we place one fermion in the lowest available state, a second particle of the same kind, with all other quantum numbers being equal, will refuse to live in that same energy state. It will occupy the next available low energy, and so forth with new identical particles added to the system.[5] The assortment of energy quantum numbers that arise from this hierarchy allows for the particles, collectively, to obey the exclusion principle.

Since electrons belong to the family of fermions, when we want to construct atoms or molecules, each additional electron added to the system must be put in at a higher value of energy. The energies of the electrons are defined so that the lowest possible state, or *ground state*, has some negative value, and each additional higher energy level is of smaller negative magnitude until the zero energy level corresponds to an electron that will no longer want to bind to

---

[5]To be clear, two fermions may have the same energy in a physical state so long as they differ in some other way, such as in their angular momentum projections.

the atomic system and is free to walk away into space. It is a well-understood principle of physics and chemistry that the electrons at higher energies (less negative, closer to zero) are responsible for chemical properties of the atoms and for chemical reactions. The Pauli exclusion principle can, therefore, be seen to be of ultimate significance in the universe in which we live: If this rule were not operative, all electrons would happily settle down to the lowest energy state of atoms and there would be no high-energy electrons involved in chemical reactions between atoms. With no chemical activity, there would be no biological activity, which in turn would result in a lifeless universe.

We should be amazed at the profound implications of this empirical rule of one particle per one state. It may come as a surprise, however, that there is no obvious way to experimentally verify the exclusion principle, although experiments have been designed to watch for violations of it (so far to no avail).[6] As it stands, it must be accepted simply as an *ad hoc* truth in order to make the quantum theory work as a description of complex matter. This should be understood as a consequence, several steps removed, of the quantum theory's origin in the planetary atomic model of Bohr.

Another remarkable feature of this modern picture of our microscopic world is that all stable matter is ultimately built from fermions and not bosons. The elementary particles out of which atoms are constructed are electrons, protons, neutrons, and— further down—the mathematical quark structures[7] in the interior of the nucleons.[8] According to the current understanding, there are two families of elementary fermions: *leptons* and *quarks*.

Electrons belong to the lepton family, along with *muon* and *tau* particles and their partner *neutrinos*. The word "lepton" has its origin in Greek where "leptos" can mean "small," "slender," or

---

[6]See, for example, C. Curceanu, *et al.*, "Test of the Pauli Exclusion Principle in the VIP-2 Underground Experiment," https://arxiv.org/abs/1705.02165v2. Despite the elaborate experimentation, it becomes clear that interpreting the results involves structural model ambiguities.

[7]We refer to quarks in this way because we hope readers who consider quarks to be real building blocks, elements of physical reality, will re-evaluate that interpretation as they read through this book. At different points, including in this chapter, we build our case that treating quarks as real particles is a quasirealist view.

[8]The *nucleons* are the proton and neutron, the particles that comprise the atomic nucleus.

"subtle."[9] The electron mass is about 1800 times less than that of the proton and is thus a relatively small-mass particle. The muon, discovered in the 1930s, is heavier[10] than the electron but still lighter than protons and neutrons. Neutrinos are particles of nearly zero mass and thus easily included in the designation. However, as the tau was discovered in 1975 and found to properly belong to the electron and muon family, the meaning of "lepton" had to undergo revision because the tau lepton has nearly twice the mass of the proton. The current understanding is that leptons are particles that do not notice the strong nuclear interaction that binds nuclei together. Leptons participate in weak interactions and, if they are charged, electromagnetism.

Muons had a rocky birth in the history of particle physics. They were first observed in cosmic rays[11] and mistaken for the hypothetical mediator of the strong nuclear interaction, the *pion*. Since the muon was found to not meet the necessary criteria of a strongly interacting particle,[12] the early reaction of the scientific community was bewilderment.[13] At the time, physicists could not see a *raison d'être* for particles other than protons, neutrons, and electrons.

The appearance of neutrinos, however, met with a very different response. Wolfgang Pauli[14] suggested the neutrino as a particle with no electric charge and almost zero mass in order to address a long-standing problem associated with an "energy crisis" in nuclear

---

[9]Leon Rosenfeld (1904–1974), a Belgian physicist, was credited with introducing the word "lepton" as a parallel to "nucleon."

[10]The muon mass is 105 MeV/c^2.

[11]While everyone seems to agree that muons were discovered in cosmic ray observations, there appears to be some confusion as to who should be credited with this feat. Anderson and Neddermeyer, Bethe, Street and Stevenson, Powell, and others are all names that have been mentioned as the discoverers.

[12]For one thing, muon's half-life of 2.2 μs is too long for strong interactions of a shorter duration than 10^{-20} seconds. More importantly, muon interactions with matter are not of the strong nuclear type. They pass through large amounts of a material medium without any interaction except leaving ionization trails.

[13]Theorist Isidor Rabi famously asked, "Who ordered that?," upon learning of the muon.

[14]Wolfgang Pauli (1900–1958) was an Austrian–American–Swiss physicist whose contributions to the rise of quantum mechanics, the neutrino hypothesis, and the Pauli exclusion principle are well known.

beta decays. Energy and momentum conservation principles dictate that if a stationary unstable particle or atomic nucleus decays into only two final products, they must each have a fixed kinetic energy. This is observed, for example, in decays via alpha particle emission. However, electron energies in beta decays vary across a smooth distribution of observed values, down to almost zero kinetic energy in some cases. This continuum of values can be easily explained if another particle of nearly zero mass is involved, sharing the kinetic energy with the electron and the nucleus that recoils after the decay. Pauli's proposed neutrino would be participating in the beta decay process, unobserved by experimental detectors. Niels Bohr was willing to entertain the possibility that energy conservation did not hold in these aspects of microscopic physics, but this was not supported experimentally and the neutrino hypothesis was also consistent with intrinsic spin considerations. The neutrino was experimentally discovered in 1956 by Clyde Cowan and Frederick Reines. We will return to a discussion of this particle later on.

The second set of fermions, the quarks,[15] are comparatively recent entrants in the elementary particle game, first theorized in 1964. Quarks are thought to be the more elementary building blocks of *hadrons*, which include protons and neutrons. It is proposed that these nucleons contain two layers of quark structure: three *valence quarks* that stand out within a background ocean containing an infinite number of *sea quarks*.

Particle physicists organize the lepton and quark families into three generations. Each generation contains four members: two leptons and two quarks, as illustrated in Table 8.1. Within each generation, the pair of quarks has a charge difference of magnitude

---

[15]The etymology of "quark" is naturally obscure but enthusiastically discussed in popular accounts of particle physics. The word was apparently borrowed from the James Joyce book *Finnegans Wake*, which contains the line "three quarks for Muster Mark!" This is a fitting story, since the quark model suggests that a proton or neutron is made of three quarks, and the word *quark* appears in German meaning curd or curd cheese, which is not a bad visual analogy to the internal consistency of nucleons within the model. Further to the point, you can find quark sold in cafeterias of German universities to the present day.

e, and the same is true for the pair of leptons.[16] For example, the electron, muon, and tau leptons all have electric charge $-e$, while the neutrinos all have zero charge, for a difference of $-e$.

Table 8.1 Quark and lepton generations

		Generation		Electric
	I	**II**	**III**	**Charge**[14]
Quarks	u up	c charm	t top	2/3
	d down	s strange	b bottom	$-1/3$
Leptons	ν_e e-neutrino	ν_μ μ-neutrino	ν_τ τ-neutrino	0
	e electron	μ muon	τ tau	-1

The first generation contains all the building blocks of atoms: electrons, and the *up* and *down* quarks that are inside protons and neutrons. The first-generation *e*-neutrino and the anti-*e*-neutrino make their appearance in weak interaction events such as beta decay or electron absorption and emission. The second- and third-generation quarks and leptons have only fleeting existence such as in the production and absorption of mesons.

8.2 Quarks and Leptons Are Really Very Different

One of the main pursuits of modern particle physics is the uncovering of nature's fundamental building blocks. This is the dream of physical reductionism. Because both mathematics and human aesthetics have a deep concern for symmetry, the fact that quarks and leptons can be grouped together in a symmetric table, as shown here, is assumed to be a gratifying outcome in the quest: At its most basic level, physical reality is anticipated to reveal profoundly

---

[16]This is the elementary charge, $e = 1.6 \times 10^{-19}$ C, which is the same magnitude of charge on the proton ($+e$) and electron ($-e$).

symmetrical order. It is striking that elementary quarks and leptons can be grouped together so elegantly, each with three generations, with doublets of two flavors in each generation, and the charge difference in every generational pair being one elementary charge. This certainly appears to hint at the bottom layer of physical reality being close at hand.

However, the symmetry and simplicity of this model are fundamentally limited and thus the quasirealist identification of quarks and neutrinos as elementary particles is only superficially supported by it. Quarks and leptons are remarkably different entities, and those differences involve fundamental asymmetries that call into question the quasirealist conclusion that we have really arrived near the bottom with the quark–lepton model of matter. There are three important reasons to see the reductionism-supporting symmetries between quarks and leptons as illusory.

First, quarks, unlike leptons, are not free entities. They are not seen, detected, or deduced from the products of interactions involving strongly interacting particles. The hypothesis of quark confinement provides a reason to understand these entities as ontologically suspect, whereas leptons live much more robust metaphysical lives.

Second, the doublet picture of quarks and leptons masks an awkward imbalance: It only includes entities that are "left handed." *Handedness* refers to how a system will appear if we change the sign of all the coordinates used to measure its position and motion in space. Mathematically, we would replace the coordinates (x, y, z) with $(-x, -y, -z)$. Physicists call this a *parity transformation* and a particle with an inherent handedness will notice. Quarks have no bias toward left or right handedness and exist in either state. From their perspective, it would have no effect on how they behave and interact if the universe was suddenly and completely flipped through a magical, global parity transformation. The world of leptons, how-ever, would notice such a transformation, because neutrinos are only ever left handed.[17] The charged leptons (electron, muon, and tau), like the quarks, also come in right-handed varieties. Hence, a complete table would include six additional quarks and charged lep-tons for the right-handed varieties, and no corresponding neutrinos.

---

[17]This is called "parity breakdown."

At this point, we think the symmetry is superficial, indeed. But there is a third, important distinction to draw: Quarks come not only in left- and right-handed varieties, but also in three colors[18] (*red*, *green*, and *blue*) and are thus found in 36 possible varieties! The leptons, on the other hand, come in only nine varieties: six left- and right-handed charged leptons and three (always left-handed) neutrinos.

Consequently, any attempt to illustrate an aesthetic symmetry between quarks and leptons is, in our view, an example of cherry-picking, leaving out properties of these entities which argue *against* the conclusion that particle physics has given us a glimpse into fundamental things. Only one of two options is credible: Either quarks and neutrinos are not, in fact, actual physical entities but effective quasiparticles whose ontological status is found at the level of mathematics; or, they are real fundamental particles in a universe that ultimately pays little heed to the kinds of symmetries that have always guided physicists' investigations.

We will now provide a few reasons to support the conclusion that quarks and neutrinos—insofar as they are currently described by theory—differ from their real charged lepton siblings in the nebulous distinction of only existing in the worldview that is defined through quasirealist interpretations.

8.3   On the Reality of Neutrinos

Only the charged leptons in Table 8.1 have a sure claim on the property of ontological physical reality. For over a century, physicists have been in a position to experimentally observe the electron

---

[18]As is often and appropriately noted in popular accounts of physics, this property of *color* has nothing whatsoever to do with the optical property of *color*. It is an unfortunate historical choice of confusing nomenclature to describe an inherent property of quarks in this way. It has been pointed out by others that, at least among non-American English speakers, one can judiciously differentiate through spelling between the optical property and the sub-atomic property. Strictly speaking, in physics, we can assign any label or name we want to a property, as long as the set of attributes are linearly independent from each other for mathematical purposes. The term "color" seems to have been inspired from the fact that white light is a mixture of distinct primary colors. For particle physics, it was necessary to assign only three linearly independent parameters and thus *red*, *green*, and *blue* were chosen.

through fairly direct means, including physical tracks left in cloud chambers. The masses of the charged leptons, especially those of the electron and muon, are measured by conventional methods.[19] The tau mass determination involves a model-dependent argument, similar to the way in which the W boson's mass is specified.

Initially, neutrinos were thought to be particles of negligible mass, but in subsequent decades many physicists naively interpreted this to mean they had *zero* mass. This assumption, however, is incompatible with the theory of relativity, which makes a significant distinction between these two scenarios in that a true zero-mass particle can never be accelerated, while a massive particle can never attain the speed of light. In the 1980s, experimental indications seemed to warrant the conclusion that neutrinos change their flavor as they fly through space, but theoretically this would require that they be of nonzero mass.

Physicists generally accepted this conclusion, but it opens up a new problem: The theory of relativity stipulates that an entity must be of zero mass to have a well-defined handedness, and neutrinos, as already mentioned, are left handed. It cannot be simultaneously true that the theory of relativity is correct *and* neutrinos have nonzero mass *and* they are left handed. The resolution of this trilemma has been to remove neutrinos two steps from reality and endow them with a complicated mathematical identity designed to effectively solve the problem. Each neutrino is now understood to be continually oscillating between all three generations. Mathematically, neutrino oscillation is described by a superposition of three unique mathematical neutrinos that must each have nonzero mass in order to sustain the oscillation. The "real" or physical neutrino is a simultaneous composite of all three mathematical neutrinos and, as a result, acquires a finite mass in the theory. This description brings us back to the problem of well-defined neutrino handedness, so the trilemma remains unresolved.

---

[19]This is true despite the fact that recent high-precision determinations of electron mass resort to model-dependent analyses. For example, see S. Sturm *et al.* "High-precision measurement of the atomic mass of the electron," *Nature*, **506** (2014): 467, who invoke "state-of-the art bound-state quantum electrodynamics" calculations to deduce the electron mass from their experimental result. These calculations involve virtual emission and reabsorption photon loops.

More interestingly, perhaps, the mathematical neutrinos are each assigned masses m_1, m_2, and m_3, corresponding to the charged lepton generations: electron, muon, and tau, respectively. Neither the exact values nor the ordering of these neutrino masses is known, although experiments do reveal that $m_2^2 - m_1^2 = 0.76\,\text{meV}^2$ and $\left|m_1^2 - m_3^2\right| = 2500\,\text{meV}^2$. This allows for two possible, very different scenarios for the relative ordering of the individual three masses. Masses m_1 and m_2 are nearly the same, with m_2 slightly larger. The pressing question (known as the "mass hierarchy problem") for the neutrino physics community is whether m_3 is greater or less than the other two. According to the neutrino physics community, resolving this problem has some importance for theories of the early universe.[20]

The trade-off in describing neutrinos in terms of these three mathematical forms is the ontological purity of the three neutrino flavors. In Table 8.1, they were understood to be three real kinds of particles that complete the symmetry among the quarks and leptons. The concept of neutrino oscillation has, however, led us to see a mathematical superposition of the original physical neutrino varieties as the *real* particle! In modern particle physics, the mathematical superposition has ontological priority over the simpler e-, μ-, and τ-neutrino states, but it comes with an unresolved problem. It is a fine example of quasirealist reasoning.

8.4 On the Reality of Quarks

Like the neutrinos, quarks are not as simple as one might think. It is misleading to think of quarks as being simply of one or another pure kind, even though they were originally conceived in this way. Our present-day quark model informs us that physical *down, strange,* and *bottom* quarks are superpositions of the original "pure" down, strange, and bottom quarks. In fact, we must write out a mathematical quark mixing matrix known as the *Kobayashi–Maskawa–Cabibbo matrix*, to account for some symmetry breakdown within particle

---

[20]Hyper-Kamiokande, "Neutrino Mass Hierarchy," http://www.hyper-k.org/en/physics/phys-hierarchy.html (accessed May 7, 2018). A figure illustrating the mass hierarchy problem may be found on that page.

physics theories.[21] In this picture, none of these three kinds of quarks are pure entities but admixtures of the other two flavors.

The fact that the down quarks are now thought to be physical superpositions, including pure strange quarks, is significant. The property of *strangeness* was originally conceived to account for a physics phenomenon that could not be easily integrated with our understanding of normal matter—the kind of matter that includes down quarks but not strange quarks. Now that down quarks are said to carry nonzero strangeness as a mathematical component (as well as bottomness), normal matter has become exotic matter after all, and the original distinction has been lost.

A different aspect of the modern quark model further erodes our confidence that particle physics can elucidate fundamental reality with its current trajectory. Because quarks are never allowed to exist as free entities flying through space, their masses must be deduced indirectly, mostly through theoretical calculations.[22] Up and down quarks are now assigned to be of two distinct types called *current quarks* and *constituent quarks*. As their name suggests, constituent quarks are the type that make up normal matter. Current quarks are considered to be the core of constituent quarks, making the latter non-elementary entities. The masses of current and constituent quarks are quite different. Up and down current quarks have masses around 2 to 5 MeV/c^2, as prescribed by theoretical calculations, while constituent up and down quarks are much heavier, each about 300 MeV/c^2.

It is not at all clear to us that the growing complexity of current models of matter is leading the particle physics community toward a

---

[21]The corresponding symmetry breakdown is known as *CP violation* and has a long, distinctive history of its own. In some literature, the physical down and strange quarks are, respectively, written d_c and s_c with the "c" subscripts indicating that they are adopted from the Cabibbo model. We also find d_w, s_w, and b_w for down, strange, and bottom quarks to indicate that they are admixtures of the Cabibbo–Kobayashi–Maskwa scheme.

[22]The most popular method is *lattice quantum chromodynamics* (lattice QCD). The QCD part refers to the quantum dynamics related to the color property of quarks. The word "lattice" arises from a method of modeling spacetime as a collection of discrete points, like a crystal lattice in solid-state physics. With experimental data as input parameters, physicists perform computer simulations using random numbers until a set of mass parameters settles out to consistent, stable values, along with other data. A 2016 review from the Particle Data Group (http:/pdg.lbl.gov) reports that these mass estimates "are not without controversy and remain under active investigation."

deeper, reductionist understanding of nature. The opposite appears to be the case, and the search for new particles or resonances has become a vocation.[23] Quarks were conceived, originally, as a device to provide order to an expanding zoo of new particles. The idea that so many particles could be described by only a small number of internal building blocks, and that a profound symmetry exists between these building-block quarks and the leptons, offered a tantalizing glimmer of hope that the reductionist theory was on the right track. However, even as recently as 1982, it was understood that the idea of real quarks was an unnecessary luxury: "It is sufficient to consider quarks as effective quasiparticles which exist inside the hadrons."[24]

Instead, we find ourselves wandering in a zoo of a different kind—not a particle zoo containing many different physical creatures, but a zoo where all the physical animals have been re-categorized as chimeras, hybrids of "pure" species, and mixed states of internal forms. The pure species, themselves, are actually mathematical idealizations without any ontological reality ascribed to them. It is no coincidence that mathematics plays the leading role in this growth of abstraction, because the community of physicists now has a long and venerated history of embracing effective ideas as the real thing. Quasirealism has become a habit.

8.5   The Gauge Bosons

So far in this analysis of elementary quanta, we have considered the fermionic quarks and leptons. As already mentioned, bosons are quantum entities with integer spin. All molecules, atoms, and nuclei that consist of an even number of particles (protons, neutrons, and electrons) are bosons but are really just complex entities built from more fundamental fermions. In fact, there are no stable elementary

---

[23]Particle physicist Steven Weinberg wrote, "A theorist today is hardly considered respectable if he or she has not introduced at least one new particle for which there is no experimental evidence" ["From Rutherford to LHC" in *100 Years of Subatomic Physics*, Eds. Ernest Henley and Stephen Ellis (Singapore: World Scientific, 2013), 5].

[24]D. Flamm and F. Schöberl, *Introduction to the Quark Model of Elementary Particles* (New York: Gordon and Breach, 1982), 16–17.

bosons in our universe that act as building blocks of matter. Mesons, which are produced in high-energy laboratories or in cosmic rays, are bosons of finite lifetimes. The other bosons that we encounter are entities that play essential roles in the interactions, and thus dynamics, of matter. These are called *gauge*[25] *bosons*, and we will point out a few reasons to see these entries in the catalogue of modern particle physics as quasireal entities.[26]

Gauge bosons arise in relativistic quantum field theories where the forces that act on and between matter are active players; the underlying force fields filling space are not just affecting the motion of matter but responding to and being modified by the matter they act upon. We can define equations of motion for the field and the matter separately. Since an equation of motion implies some entity with localization, each field is associated with a *quantum* or fundamental entity that is a kind of corporeal excitation of the field medium and conveys the field's force effects to ordinary particles. The quanta of the fields are the gauge bosons. Physicists currently identify four types of gauge boson: the photon, gluon, W, and Z.

The well-known photon is understood to be the information carrier for electromagnetism. The weak interaction involves three gauge bosons: two W bosons of electric charge $+e$ (W$^+$) and $-e$ (W$^-$), and one uncharged boson called the Z.[27] Right from the beginning, weak interaction physics was modeled along the lines of electromagnetic interactions. The most familiar examples of weak interactions are nuclear beta decays, in which an electron or positron appears along with antineutrinos or neutrinos as a radioactive parent nucleus transforms into a daughter nucleus. There are several processes in which charged leptons are created or destroyed, and other processes such as neutrino–neutrino scattering where no electric charge is explicitly involved. To mediate these different phenomena, we need to introduce both W bosons and the Z, with and without electric charge.

---

[25]The word *gauge* is used when we refer to a measuring device. Gauge bosons are so-called because they are associated with *gauge fields*, which are *gauge invariant* meaning that a measurement result should be independent of the observer's location or measuring instrument used.

[26]We return to the topic of gauge bosons in Chapter 10 with a different emphasis. There is naturally some overlap in the information and arguments here and there.

[27]W stands for "weak" and Z for "zero charge."

The weak interaction is so-called because it is mediated by a much weaker force than the strong and electromagnetic interactions, and it is considered to be of very short range. Its short range means that interacting particles need to be very close to one another for the appropriate processes to occur: less than the width of a nucleon, about 10^{-15} m. It so happens that Heisenberg's uncertainty principle suggests that short-range interactions are associated with very heavy quanta, so we expect the W and Z bosons to be quite massive.

There has been a long quest to unify all the interactions in physics. The first pair for which this was arguably achieved was the weak and electromagnetic forces, in the *electroweak theory*. This was achieved by introducing four mathematical bosons: a triplet of W bosons (W_1, W_2, and W_3) and a singlet B boson. W_3 and B are electrically neutral, while W_1 and W_2 do not have a well-defined electric charge. The physical W^+ and W^- bosons are redefined as composites of the mathematical W_1 and W_2. Finally, the Z boson and the photon are redefined as admixtures of the mathematical W_3 and B entities. It is the same kind of hybridization we have already seen with the quarks and neutrinos, and it is asserted that the mathematical construction allows photons to have zero mass while the Ws and Z acquire mass as they interact with the Higgs field.[28] Current estimates[29] suggest that a photon is about 77% electromagnetic in character and 23% weak. The Z boson is thought to have 77% weak character and 23% electromagnetic. The gauge bosons, too, have had their original identities and qualities obscured over time and to believe in them as real particles requires a quasirealist view of the relevant theories.

The next family of gauge bosons are the gluons, the quantum entities that hold things together through the strong interaction. They have integer spin of value one. As with quarks, gluons have an additional property of color, more carefully designated *color charge*, by analogy to the electric charge. Along with the three colors (red, green, blue), there are three anticolors, and according to theory, there are eight distinct gluons occurring as composite color–anticolor combinations. These combinations are properly defined in

---

[28]The Higgs field, which we address in Chapter 11, introduces yet another level of mathematical abstraction.

[29]These estimates are based on the Weinberg (weak) mixing angle. This angle is empirically determined from the ratio of masses of the W and Z bosons.

terms of orthogonal vectors within an abstract mathematical vector space. The physical gluons, then, are mathematical hybrids of pure color gluons. They do not carry a well-defined color but have only some probability of being found in a specific color–anticolor state and can thus be understood as having an intrinsic admixture of all three color states.

8.6   Summary

Throughout the discussion of elementary entities, the same challenge to realist physical convictions has appeared repeatedly: Most of the elementary entities that physicists refer to in popular accounts of the subatomic world are, in fact, formally understood in the theories to be only the mathematical mixtures of more basic entities. Thus, all gauge bosons are just part of the mathematical description of theoretical versions of one another. Since each boson has its own unique assigned properties, it is not clear how these can be retained in the mixtures and still provide meaningful distinctions with respect to the final physical constructions. Quarks and neutrinos suffer from similar confusion in their interconnectedness.

Physicists are natural scientific realists, but we have learned to trust mathematics as a reliable guide to patterns in the physical world. In the grand quest to uncover the ultimate building blocks of matter and interactions, however, we suggest that the usefulness of mathematical models has been conflated with a description of real things. We have here outlined some of the hints that this must be so. Only the electron, proton, and neutron make a clean appearance in both our experiments and theories, allowing us to point to them as, perhaps, the true elementary bodies out of which matter is constructed.

Chapter 9

What Is a Photon?

The general overview is that there is ample evidence which shows that the photon's hadron structure plays a significant role in its interactions.

—T. H. Bauer, R. D. Spital, D. R. Yennie, and F. M. Pipkin[1]

Although the photon belongs to the family of gauge bosons, the topic of chapter 10, it deserves special attention due to its unique place in the history of physics, and its surprisingly complicated character as it is understood by modern physicists, as the epigraphical quotation from Bauer *et. al.* indicates.

Soon after the four-elemental universe and concepts of atomicity were devised within ancient Greek thought, natural philosophers such as Lucretius were proposing atomic theories to explain heat, light, and magnetism. In keeping with the ancient reductionist atomic instinct, Lucretius carefully reasoned that one would find elementary objects as building blocks at the heart of even these phenomena.

Modern studies of the nature of light began with the work of two contemporaries: Huygens and Newton, each holding diametrically opposed views of the subject matter. From Rømer[2] it was known that light travels at a finite speed and not (as was previously assumed) at infinite speed. Newton and Huygens were concerned with the same

---

[1]T. H. Bauer, R. D. Spital, D. R. Yennie, and F. M. Pipkin, "The hadronic properties of the photon in high-energy interactions," *Rev. Mod. Phys.*, **50** (1978), 261–446.
[2]Ole Rømer (1644–1710) was a Danish scientist of multifaceted talents. Over 8 years, he worked at a solution to the important problem of determining a ship's longitude when out of sight of land.

From Atoms to Higgs Boson: Voyages in Quasi-Spacetime
Chary Rangacharyulu and Christopher Polachic
Copyright © 2019 Jenny Stanford Publishing Pte. Ltd.
ISBN 978-981-4800-24-4 (Hardcover), 978-0-429-02765-9 (eBook)
www.jennystanford.com

basic question about light's propagation: How does it get from point A to point B in free, unobstructed space?

Isaac Newton favored a particle-like description of light's composition. Huygens developed his ideas around wave-like propagation. These competing ideas would run side by side into the nineteenth century as physicists undertook careful experimental studies of electrical and magnetic phenomena. The dedicated and insightful work of Michael Faraday, James Clerk Maxwell, and Wilhelm Weber revealed that visible light is nothing different from electromagnetic radiation.[3] It was recognized that radiation is a form of energy, just as heat is, and in fact that heat can be transferred via radiation.

9.1 Problem of Blackbody Radiation

In the nineteenth century, as the world moved to electricity for lighting, the properties of thermal radiation were of great interest. Experiments were performed on candidate materials for high-efficiency light-emitting filaments. This led to studies of blackbody radiation: the properties of electromagnetic radiation from an object whose temperature is varied. It was found that the intensity distribution of the radiation depends solely on the temperature of the body and is independent of the material—or, more specifically, of the atomic species that make up the body.

A few decades of theoretical work went into describing the phenomenon of blackbody radiation, based on classical thermodynamics and electrodynamics. One imagines a blackbody of finite size containing an internal cavity with a small opening to allow radiation to escape. The body is held at a constant temperature, and it is thus assumed that the energy supplied to the body from its environment, and the radiation escaping from it, are at equilibrium. This means that any energy used to heat the cavity is emitted from it as radiation. This radiation was already recognized to be in the form of electromagnetic waves, so that the equilibrium state could

---

[3]The common term "electromagnetic radiation" may lead to a confusion that all radiation has some electromagnetic properties. It simply means that this type of radiation arises from sources that have electric and magnetic properties. They carry energy, linear momentum and angular momentum, and information about the changes in properties of the emitting and absorbing bodies.

be associated with *standing waves*.[4] The experimental principle for studying this phenomenon is very simple: Expose a radiation sensor to a range of wavelengths of the radiation being emitted by the body and measure the power it absorbs. At higher temperatures, the intensity has a maximum at shorter wavelengths or, equivalently, at higher frequencies.

It is quite impressive that two important deductions from these experiments—*Wien's displacement law* and the *Stefan–Boltzmann law*[5]—are still relevant to the modern cryogenic industry and fields of study such as nuclear astrophysics and cosmology. These two laws provide quantitative estimates for the dominant wavelength of radiation emitted by a body at a particular temperature, and the total power emitted across all wavelengths.

Here, in the history of physics, is perhaps the first use made of the concept of a *phase space*, which we discussed in some detail in Chapter 4. The vibrations of the wave in the cavity along each of the three spatial dimensions will be independent of one another, and this allows us to conceive of an abstract space defined by three frequency dimensions. Now we can insist on a mathematical condition that satisfies the stationary wave requirement in this frequency space and thus determine how many waves of a select frequency can exist in the cavity at a specific temperature.

What is the energy of each frequency? An attempt was made to assign the energy from the theory of gases, wherein the energy would be the same for all frequencies. It would simply be the product of the temperature of the body, T, and a well-known parameter called the Boltzmann constant, k_B, so that[6]

$$E = k_B T \tag{9.1}$$

---

[4]Standing waves originate when two or more waves are superposed as they travel in the same elastic medium. The waves are in motion through the medium, but the effect of the superposition is such that they seem to be standing still. Standing wave phenomena are associated with many everyday phenomena, including the sounds made by musical instruments.

[5]Wilhelm Wien (1864–1928) was a German physicist; Josef Stefan (1835–1893) and Ludwig Boltzmann (1844–1906) were both Austrian physicists. Boltzmann's contributions extend beyond that of blackbody radiation. A fundamental constant, known as the *Boltzmann constant*, is at the heart of the correspondence between energy and the concept of temperature.

[6]The result is from the kinetic theory of gases where the average kinetic energy of a molecule is given by $k_B T$. In the blackbody radiation theory, radiation is also assigned the same energy per wave.

It turns out that this formula only works well for long wavelengths (low frequencies). At short wavelengths (high frequencies), however, it fails so spectacularly that its failure received a name: the "ultraviolet catastrophe," wherein calculations of the body's power emission grow to infinity. This would imply that a blackbody, nearly independent of its temperature, emits an infinite amount of high-frequency radiation. Another calculation by Wien, using a semi-empirical approach, accounted for the high-frequency data but failed at low frequencies.[7]

A new model was needed, and one was proposed by Max Planck. Planck suggested that an electromagnetic wave has energy directly proportional to its frequency. His model envisions atoms or molecules vibrating as *harmonic oscillators*, like pendulums or springs, each emitting radiation in discrete quantities of the same frequency as their oscillations. The inspiration for this model most likely came from the classical *dipole oscillators* which produce electromagnetic radiation, while static charges do not. In Planck's picture, the frequency spectrum of atomic oscillation is uniquely determined by the body temperature, and thus all macroscopic entities emit an identical temperature-dependent spectrum of electromagnetic waves. The emission spectrum is continuous but modeled on a set of discretely oscillating sources.

9.2 Photoelectric Effect

While Max Planck's idea was receiving careful consideration, another phenomenon of interest had the attention of a young Albert Einstein. When light shines on a metal surface, one sometimes finds that electrons are ejected from the material. Sometimes they are not. The difference depends on the frequency of the light. There is a minimum value of frequency needed before the light is somehow able to eject the electrons. This phenomenon is different from the way, for example, that water waves of any frequency will cause erosion on a shoreline of the ocean. With the photoelectric effect,

---

[7]These theoretical treatments gave rise to a parameter called Wien's constant, as well as the Rayleigh–Jeans constant, the two parameters of importance in the cryogenics industry.

wave frequencies that are below the cut-off appear to have little effect.[8]

Einstein combined his understanding of the photoelectric effect with Planck's model of discrete emission frequencies of radiation and made one adjustment: He suggested that the light, itself, is composed of discrete bodies, or *quanta*, emitted and absorbed by the oscillating atoms. These entities, later christened *photons*, inherit discreteness from their source and are absorbed into other atoms on which they fall, to produce discrete energy changes in those receivers. Einstein's ideas constituted an ontological redefinition of the nature of light, contrasted with the Maxwellian electromagnetic wave theory in which light interferes, diffracts, and is polarized,[9] but in which no explanation is forthcoming for its emission and absorption in matter.

The concept of the photon, as a discrete energy entity or packet of light, is not very different from that of Isaac Newton, who imagined light to be made up of corpuscular bullets. But in classical Newtonian thinking, there was no association between the frequency of light and its energy. In the quantum concept, this is a chief characteristic of light: the energy of a photon is

$$E = h\nu \tag{9.2}$$

as originally given by Planck, where h is a constant of Nature. The frequency parameter for light (ν) is a continuous variable, and so is the energy that a photon may have.

9.3 Waves and Particles, Real and Virtual

Until very recently, photons occupied a unique status in particle physics. A "free" photon has zero mass, which makes the energy and momentum numerically the same value. We would recognize this

---

[8]We should note, however, that these statements are valid for photon beams of low intensities. It has been repeatedly verified that one can accelerate electrons and ions to very high energies with high-power lasers of low-energy photons. See, for example, Victor Malka, https://accelconf.web.cern.ch/accelconf/IPAC2013/papers/moybb101.pdf (accessed May 2018).

[9]Interference is a phenomenon of wave superposition. A good example is holograms, as seen on credit cards and modern bank notes. Diffraction involves the bending of waves around corners. Polarization is a phenomenon where a propagating wave oscillates along preferential directions in space. Polaroid sunglasses exploit the light filtering property of this phenomenon.

as a *kinetic momentum*. Another kind of momentum, the *canonical momentum*, is defined for a photon that is not free, but present within a material medium. As photons propagate in media, they may exert a push or pull on the material, which can be modeled as an effective mass for the photon. In this context, the canonical momentum is sensitive to the electromagnetic properties of the medium in which the photon is moving.

Maxwell's theory of electromagnetism, combined with the spacetime concepts of Einstein's relativity as a basic premise, yields a body of theory called *electrodynamics*. Early developments in quantum theory incorporated photons as waves, with quantum mechanical representations for the material bodies with which photons might interact. Photon-matter scattering was described very well by the *Klein–Nishina formula*, which treats photons as monochromatic waves of well-defined energy. Following the development of *quantum electrodynamics* (QED), theorists checked that new theory to see if it agreed with the successful Klein–Nishina formula, although the concept of radiation is ontologically very different in the two frameworks. In QED, matter particles and photons do not move in space. Rather, when a photon's position changes from point A to point B, we mathematically destroy one photon at point A and create another at point B.[10] This QED process does not obey the principle of causality and leaves the physical system in an energetically unphysical state for a short time interval—an acceptable situation if we invoke Heisenberg's uncertainty relations. An example of this kind of scenario is the well-known *Compton scattering* of a photon off a stationary electron.

Compton scattering experiments are routinely carried out in undergraduate student laboratories, where we expose a metal rod to a beam of monoenergetic gamma rays from a radioactive source, say Cesium-137. We measure the spectrum and intensity of the gamma rays that scatter from the metal by varying the detection angle with respect to the incident beam direction. The experiment involves straightforward theoretical analysis because, interestingly,

---

[10]J. J. Sakurai (1933–1982), a Japanese–American physicist, found a curious parallel in this to Hindu mythology, in which the creation and annihilation operators of QED photons are personified as Lord Brahma and Lord Shiva, two deities who create and destroy life. The product of the two operators is the *retention operator*, also known as Lord Vishnu.

the scattered photon *energy* varies with angle as if the gamma rays obey particle-like kinematics rather than the expected behavior of waves. If we ask how the *intensity* varies (that is, the number of photons that arrive at the detector at different scattering angles), we find that it follows the Klein–Nishina formula, which implies that the radiation behaves like waves.

This particle–wave dual result is consistent with the reasoning of Einstein that the transfer of energy–momentum should obey a particle-like description, while the propagation will be wave like. We should remember that for a measurement of intensity, even though we might be counting the detection of individual photons, our question concerns the outcome for a multitude of scattering events and not the result of a single photon event. This is an *ensemble* approach, and waves in Nature are always the bulk response of an ensemble of constituent entities to a disturbance in the medium. However, in terms of energy transfer, we are indeed counting each detected photon individually and asking for the energy of the photon after scattering occurs. It should be noted that we need only perform a single measurement in which we seek to know the energy of photons and their intensities in the detector. Post-experiment, we can, at our leisure, analyze the computer data. Energy information is described by the particle-like behavior, and intensities are given by wave-like behavior. It is not that photons behaved differently, but it is our questions that give different answers in this context.

To arrive at physically meaningful results, and to be in agreement with the Klein–Nishina formula, the QED theory models the Compton scattering process as a creation and annihilation of photons.[11] One aspect of the mathematics represents the loss of the incoming photon when it encounters the electron. This is illustrated in Fig. 9.1, where the electron is indicated by a straight line, and the photons by wavy lines. The electron meets the photon and absorbs it, carries on its way, and then emits a photon shortly thereafter. Notice that between the absorption and emission events, the rule that energy and momentum must be simultaneously conserved has been

---

[11]In fact, in quantum field theory nothing moves except gauge bosons. The propagation of events between two spacetime points A and B is achieved by the creation and destruction of entities and boson propagators joining the points A and B.

momentarily violated, as already mentioned. As a result, QED insists that the interacting electron becomes a virtual object during the awkward intervening time. Effectively, the electron becomes heavier than it was when real, taking on a virtual electron mass m_e^*, related to its physical mass m_e, in the following way:

$$m_e^* = \sqrt{m_e^2 + 2m_e E_\gamma} \tag{9.3}$$

where the annihilated gamma ray had energy E_γ.

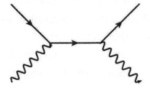

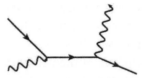

Figure 9.1 Sketch of the Compton scattering process where a photon (wavy line) scatters off an electron (solid line). The top diagram illustrates the process where a photon is first absorbed by the electron, resulting in a subsequent photon emission. The bottom illustrates that process in reverse order: An electron emits a photon to become a virtual electron which subsequently absorbs an incident photon and becomes a physical electron. Both diagrams must be included in a calculation to get a meaningful result.

This intermediate, virtual state cannot last long. According to the uncertainty relation, the amount of time available for the violation is less than about $10^{-21}/\sqrt{E_\gamma}$ seconds, with photon energies in units of megaelectronvolts. Such short times are inaccessible to laboratory experiments and thus remain without direct verification. After this momentary disassociation from reality, the unphysical electron is prodded back into physicality as a new particle, emitting a new gamma ray in the process. This emitted photon has a corresponding energy and trajectory at a specific angle that allow the original kinetic energy and momentum of the system to be fully conserved.

In addition, the mathematical formalism of QED obliges us to introduce another process into a Compton scattering event, which is understood to happen in addition to and somehow temporally parallel with the first process. In this second process, the outgoing photon is emitted from the electron *just before* the incoming photon is absorbed by it! In this situation, the electron again lives very briefly in a virtual state and has effective virtual electron mass,

$$m_e^* = \sqrt{m_e^2 - 2m_e E_\gamma} \qquad (9.4)$$

The theoretical expression for Compton scattering in QED needs to include not only these two, very different processes occurring simultaneously, but also *interference* between them. This interference is expressed mathematically as a product between the processes.

We should note, at this point, that interference is a phenomenon that takes place between waves, when their crests and troughs superpose with one another in a medium to produce regions of mutual amplitude cancellation or addition. It is exactly this wave interference that is understood to occur here, in the QED treatment of the two-process Compton scattering event. Since wave interference cannot be the outcome of the interaction of two localized particles, we are now mixing together wave and particle descriptions of photons in this model of scattering. Photons are being absorbed and emitted as if they are discrete quanta (particles), but the processes by which this happens are behaving as interfering waves. One might argue that this is not a superposition of physical waves, but of the amplitude of mathematical functions or expressions—or probability statements, perhaps, and thus only a superposition in an abstract mathematical space. If so, then this alerts us to cautiously consider how far we go in concluding that QED has provided insight into the *real physical processes* underlying Compton scattering events.

One last, important observation is appropriate: In the second process described above, where the photon emission occurs *prior to* the absorption event, we recognize that causality is being violated—the effect precedes the cause! This should be troubling to a physical realist. The only way to deal with (and thus disregard) this problem is to bury it in the mysterious Heisenberg uncertainty relation. Somehow, this relation gives prescience to the electron so that it knows what to emit before it receives the incoming quantum.

However, the problem is worse than that: We have not mentioned that there are actually more than just the two processes at work in the QED description of Compton scattering. Indeed, these two processes are only the "first order calculations," and to arrive at a mathematical result that is closer to experimental agreement, we must invoke several "higher order" processes where virtual photons are emitted and reabsorbed, and loops with particle–antiparticle pairs are employed to play different supporting roles. What should have been the simple bounce of one photon off a single electron becomes a very, very complicated collection of processes, involving all three of the four fundamental interactions of Nature (strong, weak, and electromagnetic processes). Each process is understood in QED to occur simultaneously and claim unequal but finite shares of the reality of the scattering event. Each one provides a unique answer to the question, "how did this collision take place?" A quasirealist interpretation of QED theory would have us understand that all of these processes, including those that violate the basic physical rule of causality, are justified in staking their claim, because they are part of the reality of the event—to varying degrees.

One should take a breath at this point and consider whether this is *really* happening, or, alternatively, QED is an effective mathematical treatment of complex processes we do not yet understand very deeply. If our mathematical theory requires us to accept acausal processes in our physical world, we should be very cautious. We may well be suspicious that we are being led away from a proper reductionist view of things. The quasirealist perspective is to ignore such suspicions—indeed to take pleasure in rejecting them—and naively allow the equations to define our reality.

9.4   Other Lives of Photons

Compton scattering provides an example of what should be a simple, physical interaction between light (photons) and a physical electron, but in the latest version of particle physics theory it becomes an enormously complicated quasirealist enterprise if we take it too seriously. There are other phenomena in which the photon of Planck and Einstein suffers similar reconceptualization at the hands of modern physical theories.

Electromagnetic interactions, whether between two electric charges or involving magnetic properties, are prescribed by the exchange of a virtual photon transported between the two interacting partners. A virtual photon has exotic features, which are not seen in normal photons. Consider, for example, the elastic scattering of an electron off a proton where the latter body does not absorb any of the scattering energy into its internal structure as a change in its internal energy. If we increase the energy of the incoming electron, the proton can acquire kinetic energy and be deflected at an arbitrary angle, with the scattered electron balancing the final energy and momentum of the system. In this situation, the virtual photon exchanged between the bodies has nonzero mass, just as the virtual electron of the Compton scattering process had an unphysical mass. Also, while a real photon is restricted to polarization in the transverse direction (perpendicular to its direction of motion), a virtual photon can acquire polarization along its direction of propagation as well.

In particle physics, the photon appears in many guises, depending on its energy and the context. In the early 1960s, it was found that when high-energy photons are incident on a target, such as protons, mesons are produced in the interaction. However, the experimental result shows that more mesons are produced than expected by the QED calculations for the photon–proton interaction. In response to this oddity, Sakurai proposed the *vector meson dominance model* wherein a photon can become a meson, by itself, and then interact with another particle, such as a proton, through the strong interaction. The photon has to transform into a meson without changing any conserved quantum numbers, yet acquires finite mass.[12] QED does the work to change the photon into a meson, but then the strong interaction takes over from there.

With the advent of quark models of nucleons, the vector meson dominance concept is extended to include quark–antiquark pairs in addition to lepton–antilepton pairs. Quark pairs—when energy conditions are satisfied—reappear as mesons. We may consider this picture to be looking at a deeper level than the vector meson model,

---

[12]In the early days of particle physics, there were at least three known mesons that could meet the conditions: ρ (rho), with a mass of 770 MeV/c^2; ω (omega), 782 MeV/c^2; and ϕ (phi), 1020 MeV/c^2. The ϕ meson is slightly heavier than the proton, which has a mass of 938.3 MeV/c^2.

showing an intermediate mechanism of the photon's transformation in high-energy electron–photon interactions.

We imagine an electron and photon in close proximity as the electron emits a virtual photon. The first photon simultaneously transforms into a quark–antiquark pair (a meson). The antiquark goes its own way and eventually becomes a jet of hadrons as it interacts with some medium. The virtual photon, meanwhile, somehow finds the quark and is absorbed by it. The quark is scattered by this event[13] and then, itself, becomes a second jet of hadrons. Both quark and antiquark must quickly meet their end because of the principle of quark confinement, which insists that fractional charges (thus, quarks) cannot have a free, solitary existence. Thus, the debris of the jets organize through recombination so as not to leave any fractional charges in the final state of the experiment.[14] In the laboratory, we will only observe the final state of the electron and two hadron jets. We must involve many more complicated processes if we have different experimental observables.

This illustration is a natural extension of what Sakurai and others promoted in the 1960s and 1970s. The consequence is that photon–proton interactions are no longer purely electromagnetic processes but become problems involving photon *structure*: The structure varies with the energy of the interacting bodies and is dependent on the processes being measured, but when a photon transforms into a virtual meson, it acquires mass. Does the interaction partner see a massive photon or a meson? Does it make a difference, since they are both "virtual" entities, anyway?

9.5 Photons and Electroweak Unification

Unification of forces has been an important goal and ongoing reductionist theme of particle physics. Gravitation and electromagnetism are of infinite range, meaning that each particle interacts with every other particle in the universe, but with

---

[13]Perhaps it recoils in amazement at how unlikely it is that the virtual photon should have been emitted along just the right trajectory to intercept it at just the right moment!

[14]The processes supposedly involved to make these events turn out "just so" suggest a surprisingly teleological aspect to the subatomic world. Aristotle and Thomas Aquinas might have been very comfortable with QED.

decreased influence as separation distance grows. Nonetheless, the influence is always finite. The weak interaction was modeled on the electromagnetic interaction, but theorists had to account for the fact that the weak interaction has only a very localized effect.

The process of beta decay, where a neutron transforms into a proton, electron, and antineutrino under the causal influence of the weak interaction, was at one time considered to be a *contact interaction*: The neutron instantaneously disappears at a point in space and is replaced by the other three bodies. The contact interaction picture was later relaxed, and a *boson exchange mechanism* was invoked instead, in line with the idea behind the cause of gravitation and electromagnetic forces.

Some clarification of exchange bosons may be helpful here. A photon is an exchange particle to the extent that it inherits its properties from its source conditions (the quantum numbers of the states involved in giving rise to its existence) and then imparts these properties to the object with which it subsequently interacts. The receiving object only knows the properties of the photon; it has no direct information about the parent or source state. Thus, the exchange mechanism is not a direct transaction between the source and receiving particle but is mediated by the exchange boson that acts as a messenger: The receiving object is affected by the boson's properties and responds accordingly.

Any question of interactions between particles, then, naturally concerns the properties of the messengers, and there arose a sense of obligation to unify the electromagnetic and weak interactions into one theoretical framework. An experiment with neutrinos confirmed that a neutral (zero-charge) boson (called Z) along with two bosons of charge $\pm e$ (called W) are required to act as mediators for the weak interaction. The positive and negative W bosons are a particle–antiparticle pair and thus have identical masses.[15]

---

[15]The requirement that a particle and its antiparticle must have the same mass is based on a very fundamental symmetry of particle physics, known as *CPT theorem*. It says that if we do a time reflection (change the flow of time in equations from positive to negative or vice versa), as well as change a particle into its antiparticle and reverse the positive and negative axes of our spatial coordinate system, the result is the same as not having changed anything. With this trio of simultaneous changes, we simply recover our original system. If this theorem is proved wrong, seemingly all field theories will be invalid. So far, no experiment disproves this theorem.

The neutrino experiment also provided a ratio between the W and Z boson masses in terms of a parameter called the *Weinberg angle*. The Weinberg angle is a mathematical construct[16] that makes the photon and Z boson superpositions of something mathematically more fundamental. At that point, the physical photon and Z boson are no longer considered fundamental entities in electroweak theory, but mathematical expressions of two new mathematical bosons. They are mixed together in this new identity, and thus the massless electromagnetic photon now acquires a finite aspect of the Z boson within its mathematical self. The photon has about 23% of its identity contributed from a Z component.

This is not a negligible admixture of the weak interaction mediator, so why is there not more confusion of weak and electromagnetic forces in everyday natural events? The answer is that the weak interaction does not normally play much of a role in physical phenomena at low-energy scales, so the electromagnetic contribution can have exclusive reign while the Z component of the photon effectively hides, unnoticed.[17] And how can the physical photon remain massless when part of its identity is made up of enormously massive Z? Just as spontaneous symmetry breaking gives particles their respective masses through the mediation of the Higgs boson (as discussed in Chapter 11), the photon is understood to be immune to this breaking and thus remains massless in keeping with the requirements of the Standard Model of particle physics.

9.6 Are Photons Phoenixes?

Experiments show that as light traverses a medium, its intensity drops off exponentially, a process called attenuation. Suppose we let a beam of photons travel through a medium to observe its passage over a certain length of time. The loss of intensity relates to the ongoing interaction of a photon with the constituents of the medium, through absorption. At any given moment in its journey

---

[16]This angle is no more an angle in physical space than the phase angle of impedance in an electric circuit (see Chapter 4).

[17]However, the large decay widths of Z bosons, and their decay into hadrons, make this assertion that the Z boson participates solely in weak interactions questionable.

through the medium, the probability of a photon being absorbed is independent of its history—how far it has traveled in the medium. A photon's instantaneous probability of absorption does not increase just because it has, thus far, managed to avoid being absorbed. The loss of intensity of light shining in a material is, therefore, independent of the history of the photons. Also, because the photons that continue to survive are not being absorbed, there is no change in their energy. This is unlike the passage of charged particles such as protons or alpha particles, which gradually and continuously lose energy throughout their journey through ongoing interaction with the medium.[18]

This implies that the existence of a photon in a medium is a one-shot process. They travel an arbitrary distance without interaction, retaining their physical properties, including energy and momentum, unchanged. But if they do interact, they are lost from the main beam. QED suggests that new photons will emerge from these interactions, and overall the result will satisfy all conservation laws. But the new photon that is emitted is not the original quantum that was absorbed; it has new energy and momentum values and arises out of the internal processes of the atomic dynamics. These photons are like the mythological phoenix, arising out of the ashes of death.[19] In the very act of seeing or detecting light, we annihilate the photons that provided our sensation or signal and they are no more.[20] Between creation and destruction, then, we can never have experimental knowledge of a photon's state, for every measurement will involve an end to that photon we wanted to observe.

---

[18]Incidentally, these distinctive behaviors of photons versus charged particles are being exploited in radiation therapy and medical imaging.

[19]Lest one quibble here and suggest that phoenixes experience a resurrection of their personal identity and are thus the antithesis of the very point we are making about photons (annihilation), we will simply argue that some ancient authors, such as Clement of Rome, understood the mythological Greek phoenix rebirth to be a means of reproduction: The offspring of the phoenix arises from the parent's ashes. Perhaps, then, we should more cautiously compare our photons to "Clementine phoenixes," since confusion on this point is bound to be widespread, given the precarious modern state of phoenix conservation efforts.

[20]This observation inspired one of us to write a short epitaph for photons, shared in a conference talk and included as an appendix, lamenting that we should not celebrate their life but only mark their demise.

9.7 Finally, What Are Photons?

We thus arrive at a complicated picture of our photon, the quantum of electromagnetic radiation that Newton conceived of as a corpuscle and Huygens as a simple wave. Our mathematical treatments have it traveling as a wave. But it behaves like a particle when it exchanges energy or momentum with its environment. It may convert itself into various particle–antiparticle pairs such as electron–positron pairs or other leptons. It can transform into vector mesons or quark–antiquark pairs, or even gluon loops. Above all, though, about 23% of the time (or all the time, but with only 23% conviction?), it is a neutral weak boson. It is of zero mass when it moves through free space, but it acquires mass as it moves in material media.[21]

In a nutshell, we may attribute many incarnations and reincarnations to a photon, each having a different probability of occurrence throughout the photon's nebulous voyage in (quasi) spacetime. Figure 9.2 is a diagrammatic representation of some of these possibilities.

Furthermore, we emphasize that a photon does not have an independent existence in physical reality, and this may be somewhere at the heart of the problem, since they are normally described among physicists in terms of independent reality. Photons gain their properties from the entities involved in their emission. The properties of light are modified by the medium it moves within, although the individual photon remains unmodified until it meets its demise through absorption. We are denied the opportunity to perform any observations on the life of a photon between its birth and its death because the very act of observation will entail absorption and demise, so our picture of a physical photon is limited to the information about its beginning and ending only, as seen through the changes in its interacting partners' properties.

---

[21]Since quantum field theories require that the vacuum is an active medium with much in common with the ancient idea of *plenum*, "free space" is only an idealization from the perspective of the most modern physics. This raises the concern that there may be no such thing at all as a "free photon" traveling at a constant speed c. What then of the consequences of special relativity that relate to this remarkable speed limit of the universe?

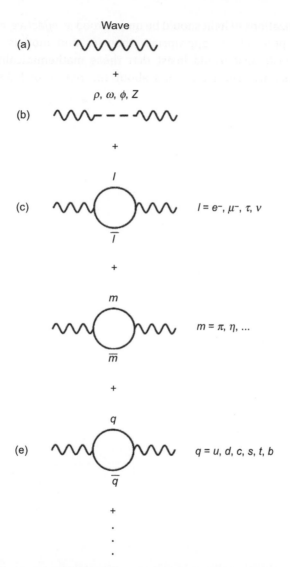

Figure 9.2 The multifaceted life of a photon can take many forms: (a) a classical wave; (b) a virtual vector or gauge boson; (c) a virtual lepton–antilepton pair; (d) a virtual meson–antimeson pair; (e) a virtual quark–antiquark pair; + innumerable others...

None of these statements about photons are groundless, or without motivation, but that is not the same as saying they tell us something true about physical reality. Many, or all, of these characterizations of light should be understood as *effective*, confined to the context of the appropriate theories and models, so that only a quasirealist would insist that these mathematically dense conclusions are true statements about the reality of light in our universe.

Chapter 10

Symmetries, Conservation Laws, and Gauge Bosons

Symmetrie, ob man ihre Bedeutung weit oder eng faßt, ist eine Idee, vermöge derer der Mensch durch die Jahrtausende seiner Geschichte versucht hat, Ordnung, Schönheit und Vollkommenheit zu begreifen und zu schaffen.[1]

Symmetry, as wide or as narrow as you may define its meaning, is one idea by which man through the ages has tried to comprehend and create order, beauty, and perfection.

—Hermann Weyl

10.1 Symmetry and Gauge

The mathematician Hermann Weyl correctly highlights the abstract concept of *symmetry* as a crucial and intuitive property of natural systems that endurably reappears throughout the history of natural philosophy and science. Symmetry reliably guides practitioners towards increasingly accurate models of physical reality. It plays a very important role in physics, as it does in many other areas of human activity such as art, music, and literature. Symmetries in particle physics are related to some specialized terms that will be encountered early on and often by students of the discipline: *gauge*

---

[1]Hermann Weyl, *Symmetrie* (Stuttgart: Birkhäuser, 1955), 13.

From Atoms to Higgs Boson: Voyages in Quasi-Spacetime
Chary Rangacharyulu and Christopher Polachic
Copyright © 2019 Jenny Stanford Publishing Pte. Ltd.
ISBN 978-981-4800-24-4 (Hardcover), 978-0-429-02765-9 (eBook)
www.jennystanford.com

fields, gauge bosons, gauge transformations, gauge symmetries, and *gauge invariance*. It is worthwhile to briefly review these terms and their meanings before commenting in this chapter on the influence of quasirealist thinking within the context of symmetry and, subsequently, on the ontological status of gauge bosons.

In typical usage, a *gauge* is a device to measure something, such as a fuel or pressure gauge. One may then interpret the term "gauge invariance" to mean, quite generally, that the result of a measurement is independent of the measuring device used. In the specific context of physics, we take it to mean that the result of a theoretical formulation should be independent of the observer's frame of reference or point of view. A physical theory is required to explicitly exhibit this constancy. Hermann Weyl coined the German term "eichinvarianz," which is now translated as gauge invariance. In German, the prefix *eich-* means "calibration," which suggests that the system described by a physical theory is invariant under an arbitrary change in the coordinate system in which it lives. In other words, the physical system is entirely unaffected by the choices we make as to where the origin (zero point) of our coordinate system lies relative to the system or the scale of units used in calibrating our measurements of any parameter such as position or duration.

It was known even from the early eras of Galileo and Newton that the physical laws governing an event or phenomenon must appear to be the same for all observers. This implies that the mathematical formulae describing physical laws should also look the same for everyone. This will certainly include the specific case where we consider the motion of charged particles in electromagnetic fields or the propagation of light. Here we resort to Maxwell's formalism of *fields* and *potentials*. The fields are linked directly to *forces*, from which we can deduce observable effects. The field of a force (such as an electric or magnetic field) can be thought of as a kind of map in space, showing the influence of the force on any body that is sensitive to it. For every known field, one can define an infinite number of unique potentials, which are themselves a kind of map of the potential energy associated with bodies exerting and feeling the force in question. These potentials differ from one another by some arbitrary parameter that does not, in itself, affect the physical influence of the force/field on its environment. That is to say, for a specific potential, the corresponding field is uniquely determined, but the reverse cannot be assumed.

In terms of ontological status, forces enter our experience directly as felt pushes and pulls with obvious effects on their environment. Fields have a somewhat less certain status as aspects of physical reality, although it is convenient to consider them as having a real existence, and not uncommon to even ascribe to fields an ontologically prior status as the true *cause* of forces. Potentials, however, are another matter. The fact that a unique field is associated with an infinite number of unique potentials renders potentials ontologically suspect. Indeed, Maxwell found it desirable to remove potentials from equations, considering them to not possess any real physical significance.[2]

A gauge transformation involves a change in the potentials, with no subsequent effect on the observables of a physical system. We cannot observe the physical effects of a gauge transformation because there are no such effects. If we manage to perform these transformations without affecting the fields, and thus the outcomes of a measurement, the system is said to be "gauge invariant," and the corresponding field is called a "gauge field."

10.2 Gauge Invariance and Electromagnetism

We run into a difficulty, however, when we try to impose gauge invariance in the context of electromagnetic fields and light propagation. Maxwell's equations prescribe a preferred frame of reference for the vacuum in which light moves at a constant speed, independent of the motion of the light source. Einstein's special relativity[3] partially alleviates this problem for the special case of observers moving with constant velocity with respect to one another,[4] retaining gauge invariance through a formalism of *global transformations* that exhibit *global symmetries*. They amount to an abstract redefinition of the structure of space and time in terms of imaginary numbers. The simplicity gained through special relativity's transformations is lost, however, in general relativity.

---

[2]See, for example, A. C. T. Wu and Chen Ning Yang, "Evolution of the concept of the vector potential in the description of fundamental interactions," *International Journal of Modern Physics A*, **21**, 16 (2006): 3235–3277.

[3]We discussed special relativity in Chapter 3, explaining its relation to quasirealist interpretations in physics: Metaphysical acceptance of special relativity is accompanied by the cost of making the spacetime concept ambiguous.

[4]That is, observers who occupy *inertial reference frames*.

There, we must define a distinct set of parameters for each localized group of points in one reference frame, with respect to a localized group of points in a second frame.

Ultimately, we insist on finding mathematical formulations that will exhibit local symmetry. To this end, the formulations must not be explicitly dependent on any particular potentials. In the case of a quantum theory of electrodynamics, however, which describes the influence of electromagnetic fields on quantum mechanical systems, our formalism mathematically incorporates field effects directly by including a potential parameter in the kinematical term of the equation for the overall energy of a system. The observable dynamics of the system are mixed up with the field potential. The momentum of a quantum mechanical particle, such as a proton, in the presence of an electromagnetic field comprises both a kinetic momentum (akin to the classical $p = mv$), and an additional component given by the product of the charge, e, with the field potential, yielding a term eA, where A is the *vector potential*.[5] The observable motion of the particle is thus directly linked to a non-physical potential, and it appears that we no longer have a clear demarcation between physical entities and the interactions that affect them. Let us recall that quantum mechanics also allows for so-called *zitterbewegung* (wavy motion) of a particle in the absence of any external force, which has already taken us far away, in a conceptual sense, from the notion of a free particle living according to Newton's laws. The melding of momentum with mathematical potential terms in quantum electrodynamics is another large step out of the Newtonian (and Galilean) world.

10.3 Symmetry and Isospin

Generally speaking, an invariance principle or preserved symmetry is related to a *conservation law* or a *selection rule*. Emmy Noether[6]

---

[5]Interestingly, although Maxwell considered the potential to be no more than a flexible mathematical convenience, he recognized it was related to a momentum due to external fields.

[6]Amalie Emmy Noether (1882–1935) was a German mathematical physicist. She migrated to the United States due to the political situation in Nazi Germany. Noether's elegant theorem of symmetry and conservation principles is easily accessible to an undergraduate physics student and can be found in any introductory book of quantum mechanics.

proved a very simple yet powerful theorem on how to identify conserved quantities in terms of symmetries for both classical and quantum systems. Accordingly, we can examine both classical and quantum mechanical symmetries, whether abstract or physical, and evaluate the validity of a corresponding conservation law.

A good example of this is the *isospin symmetry*, which, from Heisenberg's prescription, treats neutrons and protons as a doublet of a hypothetical particle called the *nucleon*. In the abstract mathematical electric charge space,[7] we can consider the transformation of a proton to a neutron as a rotation from one projection of a vector to another. Recall that gauge invariance suggests that changes in the definitions of coordinate systems (like rotations) do not affect the physics of a system. It should be noted that there is no physical process that just "rotates" a proton to a neutron or vice versa; anything of this sort would violate the conservation of electric charge, so the idea must be held on strictly hypothetical terms.

To fully embrace the isospin symmetry between protons and neutrons, one must assume that electromagnetic interactions play no role in the physical context, which of course they do. So the isospin treatment should be viewed as a clever but abstract mathematical tool that does not provide us with information about a *real* correspondence between protons and neutrons. To take isospin symmetry too seriously would be a *quasirealist* move. After all, the physical proton–neutron transformation does not happen in isolation but is accompanied by other interaction partners being created or destroyed.

Keeping in mind this caution, the isospin trick has been extended to other members of particle families with many spectacularly useful results for mathematical ease and the identification of symmetries. In quantum chromodynamics, the *color* degree of freedom is an abstract symmetry and a simple extension of isospin. The Higgs field, out of which the Higgs boson appears, also has theoretical connection to this mathematical notion that originated with Heisenberg. Abstract symmetries have a powerful role in the practice of particle physics, and isospin has provided a way forward to invent some new ones.

---

[7]See our comments on mathematical spaces and quasirealism in Chapter 4.

10.4   Mixing of Matter and Interactions

Recall that the treatment of the electromagnetic field potential in quantum electrodynamics results in a mixing of particles and the interactions that act on and by them via the fields. An interesting feature of relativistic quantum field theories is the coexistence of two Lagrangian components in one theoretical expression: Both a force field and a matter field (which is affected by the force) are active players in the model, and we can demand the equations of motion for the field and matter separately. As an equation of motion implies a moving entity, each field in this formulation is associated with a *quantum*, which refers to a material particle or entity that is supposed to be the basic localization of a field. These quanta[8] must be bosons (having integer intrinsic spin) in order to conserve spin overall. Once we agree that there are bosons living as quanta in our universe, the next question should be, "What are their masses?" Indeed, gauge invariance requires that these bosons must be of zero mass.

Photons, the electromagnetic quanta, were the first kind of gauge boson to be identified as such. The electromagnetic field quantization is compatible only with zero-mass photons. If we introduce a photon with finite mass into the theory, the symmetry under the gauge transformation is broken. So we are quite satisfied with the zero-mass condition in this case.

The theory describing the physics of the weak interaction was modeled, from the start, along the lines of the electromagnetic interaction. Just as there is a quantum for electromagnetism, one is inspired to posit a similar kind of bosonic entity for the weak interaction. In Nature, we find several scenarios in which it appears that charged leptons (electrons, muons, and the tau) are created and destroyed. We are also aware of neutrino–neutrino scattering. The most familiar examples are nuclear beta decays, where an electron or positron appears along with neutrinos when a proton in a nucleus effectively converts to a neutron or a neutron to a proton. The interaction field that governs this kind of event would require *both* charged and uncharged gauge bosons as its quanta. This is unlike the situation for the electromagnetic interaction, which acts between electrically charged bodies or electromagnetic fields acting

---

[8]Quanta is the plural of quantum.

on charged bodies, for which only a zero-charge boson was found to be sufficient.

The weak interaction field thus requires a set of three gauge bosons, labeled W⁺, W⁻, and Z, with electric charge +e, –e, and zero, respectively. These are understood to be responsible for mediating the physics of the interaction, which is much weaker than the strong and electromagnetic interactions and is also considered to be of very short range: The interacting partners must be very close to one another for the weak interaction processes to occur. If we invoke Heisenberg's uncertainty principle as a guide to the relationship between interaction range and the mass of the associated gauge boson, it is possible to conclude that these bosons may be quite heavy since their influence is manifest on a length scale of less than the size of a nucleon (about 10^{-15} m).[9]

The development of the weak interaction theory was merely a step in a long quest to unify all the interactions in Nature.[10] The first pair of interactions for which this was arguably achieved was these two forces of Nature, combined into the *electroweak theory*. This unification allows one to estimate the masses of the W and Z bosons, with the following assumptions: First, the weak charge is of the same magnitude as the electric charge; second, the neutral Z boson and the photon are admixtures of more basic mathematical entities.

To explain the meaning of the second assumption, we must venture into the realm of some abstract mathematics and should thus immediately be on guard against quasirealism—that is, taking the results seriously as physically informative. Let us assume that there is a vector representing the weak bosons, which has three projections onto three independent axes. We will call these projections W_1, W_2, and W_3, and the trio of them a *weak triplet*. We also have a *singlet* "zero projection" called B. Projections W_3 and B are both of zero electric charge, while W_1 and W_2 are ill-defined in

---

[9]The association of heavy bosons with a short distance of interaction is not strictly required in all cases. For example, gluons—the mediators of strong interactions among quarks and antiquarks, which act in the interior of hadrons—are assigned zero mass.

[10]In the second chapter of this book, we explain how this quest for a theory of everything provides an important stimulus for physicists to uncritically adopt quasirealism as a worldview for understanding abstract mathematical ideas in their discipline.

this sense, being mathematical combinations (*superpositions*) of +e and –e charges.

The prescription is that the "physical" bosons W⁺ and W⁻ are, themselves, mathematical combinations, or superpositions, of the vector projections W_1 and W_2. The zero-charge components B and W_3 mix together in different ways to yield the Z boson and photon.[11]

We may write the four resulting gauge bosons of electroweak theory as follows:

$$W^\pm = \frac{1}{\sqrt{2}}\left[W_1 \pm W_2\right] \tag{10.1}$$

$$Z = \alpha B + \sqrt{1-\alpha^2}\,W_3 \tag{10.2}$$

$$\text{photon} = -\alpha W_3 + \sqrt{1-\alpha^2}\,B \tag{10.3}$$

In this formalism, none of the physical gauge bosons are mathematically basic. They are all combinations of four more basic mathematical entities, and this raises a question of ontological priority: Which are the true gauge bosons of Nature?

There is also a curious problem of inconsistency in the masses of the bosons. According to current estimates,[12] a photon is about 77% electromagnetic quantum and 23% weak quantum. The Z boson has the opposite composition. It is asserted that this mathematical construction endows photons with zero mass, despite the fact that the W and Z bosons acquire mass as they interact with the Higgs field. The Higgs field is, by the way, a higher level abstraction of mathematical fields. Taking the gauge bosons seriously as real particles is becoming increasingly difficult without a blind dive into quasirealist metaphysics.

The latest literature[13] tells us that the mass of the W boson varieties is 80.387 GeV/c^2, and that of the Z is 91.187 GeV/c^2. We

---

[11]As taught in undergraduate quantum mechanics angular momentum algebra, W_3 is an *eigenstate* of zero projection, while W_1 and W_2 are not *eigenstates*, but can be combined to make raising and lowering operators. In the electroweak theory, W⁺ and W⁻ are the raising and lowering operators corresponding to +e and –e electric charges.

[12]These estimates are based on the Weinberg (weak) mixing angle. This abstract mathematical angle is empirically determined from the ratio of the masses of the W and Z bosons.

[13]The best resource for this is the Particle Data Group website, http://www.pdg.lbl.gov.

cannot leave this discussion without remarking on an amazing sleight of hand performed by Nature in the process of radioactive decays, if our understanding of these masses is physically reliable. The role of these bosons as mediators of the commonly encountered radioactive decays, such as beta decays, implies that either the decaying nuclei are borrowing energy from somewhere amounting to about 80 or 90 times that of a proton mass, or that the W and Z mediating the decay are exceptionally different from the supposedly real ones of the theory (not to be confused with their mathematical building blocks). This is because the energy emitted in the beta decay process is always less than a few MeV, more than 1000 times lighter than the energy required to create these intermediary bosons. Of course, we resort to Heisenberg's uncertainty principle to provide the missing mass and thus imply that these decay processes occur over very short distances in very short time intervals. Nonetheless, the insistence that such an enormous quantity of energy fluctuation is truly occurring out of nothing, in order to call into existence a mathematically defined entity, is quite unsettling.[14]

Also disconcerting is the observation that only about 10% of the Z boson decay events yields weakly interacting electron, muon, and tau leptons and their neutrinos. About 20% is listed in the literature as "invisible" decays, likely occurring as neutrino–antineutrino pairs. The remaining 70% of Z-mediated decay events results in strongly interacting hadrons! We think the inconsistency of this detail should be an adequate cause for philosophical reflection on any reality ascribed to these processes. Should we not question why this—if a physical, real boson whose sole *raison d'être* is to mediate weak interactions—chooses to decay through strong interaction channels, in a short time interval (less than 10^{-25} seconds) which is of a timescale (according to present-day theories) characteristic of the strong interaction?

Theoretical models, attempting to account for the short lifetime of the Z, invoke several complicated processes where the boson emits various mesons, which are interacting by strong and electromagnetic

---

[14]Somewhere we have encountered a textbook analogy of a bank manager or teller taking money from bank vaults during the hours when the banks are closed, gambling with the money, and winning a fortune for himself to keep, while returning the borrowed funds before the next business day and not getting caught. An analogy to label Nature as a thief is troubling. And what exactly, in Nature, plays the role of the casino? This question may be especially pertinent when, according to Einstein, God does not play dice with the universe.

interactions in addition to the weak interaction to which the Z was originally designed. There are many other processes besides those. They can all be illustrated using the familiar approach of drawing *Feynman diagrams*. Some relevant examples, illustrating how the Z boson manifests in its short life and transforms into other entities, are shown in Fig. 10.1 as *loop diagrams*. We can see that a Z boson can become a quark–antiquark pair or a virtual[15] W$^+$-W$^-$ boson pair (its weak interaction partners), or all kinds of other exotic particle life, including coupling with the Higgs boson.

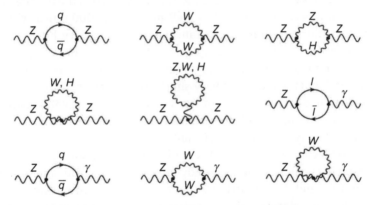

Figure 10.1 Feynman diagrams illustrating some of the expected transformations of the Z boson. The top three diagrams from left to right depict the Z transforming itself into a bottom quark–antiquark (q) pair, a pair of W bosons, and another virtual Z plus a Higgs boson (H). Reading each diagram from left to right as the flow of time, not space, the idea is that the Z fleets out of existence for a brief moment in time and then reappears. The first two diagrams of the middle row show more complicated interior structures forming. The last four diagrams represent a Z decaying into a photon with intermediate states of lepton–antilepton, quark–antiquark, and W character.

Another family of gauge bosons is the gluons, which describe strong interactions, also of short range. Gluons have an intrinsic

---

[15]Virtual particles are assigned the same spin, parity, and other properties as their counterpart real particles. However, they do not occur as free entities in the laboratory. In Feynman diagrams, one can recognize virtual particles as not having an open end but occurring between two points at which other particles meet or depart from them. They can also appear as closed loops at one point or between two points, flashing in and out of existence, never visible in any experiment. These fleeting events must satisfy momentum and charge conservation but not necessarily energy conservation.

spin of value one. Similar to the electric charge in electromagnetism, physicists conceived of a hypothetical charge for the strong interaction and gave it the name "color charge."[16] According to the theory, there are eight distinct gluons occurring as composite color–anticolor combinations, which means individual gluons do not have well-defined color or anticolor properties, but only some probability of being found in a specific color–anticolor state. The mathematical form of these eight gluons, in terms of their color charge, is the following:

$$(r\bar{b} + b\bar{r})/\sqrt{2} \qquad\qquad -i(r\bar{b} - b\bar{r})/\sqrt{2}$$

$$(r\bar{g} + g\bar{r})/\sqrt{2} \qquad\qquad -i(r\bar{g} - g\bar{r})/\sqrt{2}$$

$$(g\bar{b} + b\bar{g})/\sqrt{2} \qquad\qquad -i(-g\bar{b} + b\bar{g})/\sqrt{2}$$

$$(r\bar{r} - b\bar{b})/\sqrt{2} \qquad\qquad (r\bar{r} + b\bar{b} - 2g\bar{g})/\sqrt{6}$$

Here the letters correspond to specific color charges in this way: *r* for *red*, *b* for *blue*, *g* for *green*, and a bar overtop indicates the corresponding anticolor. The gluons, then, are ontologically similar to the other bosons we have discussed in being mathematical composites of more fundamental non-physical entities.

All the gluons have zero mass, mediating the strong interaction over very short distances of less than the nucleon radius (about 0.8 femtometers). If we remember the Heisenberg uncertainty principle, once again, these quanta would be expected to have very large mass, like the W and Z bosons, perhaps several hundred MeV/c^2 or greater. However, particle physicists have had to assign them zero mass, like the photon, to satisfy the requirements of gauge invariance symmetry. This stands in contrast to the theoretical adjudication that W and Z bosons of the weak interaction are allowed to have mass and break this symmetry.

---

[16]It might be disappointing to some that physicists settle on unappealing, everyday vocabulary for these particle names. The term "gluon" is indicative of something that holds things together, so the name has some utility. The term "color" is, unfortunately, without any relevance to the everyday use of the word colour/color that we use to describe what our eyes see. There is no connection between the two kinds of colour/color.

10.5   Conclusion

From the perspective of theoretical sophistication, the gauge bosons are an impressive and useful species of quasiparticle mirroring the underlying symmetries of Nature. We see, however, that none of these are mathematically pure entities. In the case of the charged W bosons, the mathematically elementary W_1 and W_2 do not possess a unique, and thus *physical*, electric charge. In the case of the neutral Z and photon, neither of them is uniquely weak or electromagnetic quantum. Electroweak unification is achieved by blurring the boundaries of these two interactions, along with the purity and simplicity of the quanta. This seems to us to represent a step away from the reductionist vision. That kind of retreat is characteristic of the fine details of quasirealism.

As a final observation, the Higgs boson is unique in that its sole mandate is to give masses to other quanta. It has not been assigned the status of gauge boson but is uniquely called a *scalar boson*. We devote a detailed discussion of the Higgs in the next and final chapter, where we explore this particle's relationship with our concept of quasirealism.

Chapter 11

Higgs Boson

We are all agreed that your theory is crazy. The question that divides us is whether it is crazy enough to have a chance of being correct.

—Attributed to Niels Bohr by Freeman Dyson, "Innovation in Physics" *Scientific American*, **199**, 3 (September 1958), 74–82

11.1   Knowing What We Cannot See

How do we know that what we see[1] is what we think we see in particle physics? Physicists make experimental arrangements to affect the production and detection of subatomic entities. The presence or absence of these entities must be inferred from the evidence, without the benefit of first-hand sensory data. The situation is rather like the case of a homicide investigation, where a forensic investigator gathers data and a prosecutor builds a reasonable case by post-mortem analysis, to persuade a jury beyond reasonable doubt. In particle physics, however, the physicist plays the combined role of investigator, prosecutor, and jury throughout the investigation and trial.

For many kinds of particles, such as protons, pions, and kaons, we can determine the presence and properties of each individual

---

[1]We use the term "see" in the most general sense, not limited to human sensory organ perceptions. Those deductions based on instrumental responses, and inferred from our physics understanding, qualify as "seeing" in this sense.

From Atoms to Higgs Boson: Voyages in Quasi-Spacetime
Chary Rangacharyulu and Christopher Polachic
Copyright © 2019 Jenny Stanford Publishing Pte. Ltd.
ISBN 978-981-4800-24-4 (Hardcover), 978-0-429-02765-9 (eBook)
www.jennystanford.com

body from its measured speed and trajectory in an electromagnetic field. For heavier mesons, W bosons, and quarks, however, we do not see the particle's individual signature in the apparatus. In such cases, we must resort to *event reconstruction* from the signatures of many decay products and apply *model arguments*—an appeal to our best theory to provide guidance for our interpretation of the data— to identify the particle and assess its properties, such as the mass. Event reconstruction involves analysis of all the bodies that enter our detectors, or some specified subset of detected entities. We then work backward using conservation law calculations to determine their parentage.[2] When we use model arguments for assistance in this, we may appear to risk the logical fallacy of *begging the question*: assuming the very thing to be true that our experiment is supposed to test. Nonetheless, in recent times, the criteria for positive particle identification have become increasingly and necessarily influenced by model-dependent reasoning.

11.2   Searching for the Higgs Boson

It took nearly half a century for the Higgs boson to be hunted down by particle physicists. As is now well known even among non-specialists, the Higgs particle is understood to play an important role in fundamental physics, providing the property of mass to other entities that *couple* to the *Higgs field*, which is the theoretical, mathematical space-pervading background medium in which Higgs bosons arise and live as mathematical excitations of its mathematical substance. At first, physicists had no idea what the Higgs mass would be, and no clue how to go about finding it, because no one knew what the decay products of a Higgs particle should look like in their detectors. However, some things *were* known about the particle: It is of zero electric charge, zero spin, and positive parity, a so-called *scalar particle*.[3]

As a first guess, the Higgs might be produced in any particle–antiparticle collisions with total energy greater than its mass. It might also decay into any particle–antiparticle pair with masses

---

[2]In Chapter 5 on Mass, we have provided an example of an event reconstruction using invariant mass.

[3]In the terminology of particle physics, a *scalar particle* is one which has no sense of direction. A zero spin and positive parity system is isotropic, and changing the sign of its coordinates in space leaves the particle properties unchanged. Technically speaking, the wave function does not change under such a transformation.

less than its own. However, we can imagine that other particles are produced along with it, and that it might produce more than just particle–antiparticle pairs. Photons are their own antiparticles, and thus a pair of photons is always a possibility. However, the models suggest that photons would not couple to the Higgs, since they have zero mass. This makes photon involvement, whether real or virtual, seem less likely than other possibilities.

Speculative values for the mass of the Higgs particle ranged from almost zero to several gigaelectronvolts. In a beautiful article, Ellis, Gaillard and Nanopoulos[4] critically and cautiously examined several possible modes of production and decay channels for the Higgs. This paper illuminates the method of thinking, logical rigor, and cautious optimism of good particle physicists. They suggested several indirect methods of observing the boson, such as through the spatial distribution of nuclear reaction products and reconstructions of missing mass. They also identified two-photon decays as a possible context for discovery, if the Higgs had a small mass equivalent to an energy on the order of a few megaelectronvolts (MeV), albeit as a result of an unlikely process involving intermediate virtual particles. Not long after, Higgs boson searches became the exclusive endeavor of high-energy physics. Dedicated experimentation, as well as the confirmation of the W and Z bosons and the top quark, limited the mass of the Higgs to less than 1 TeV, within the reach of present-day particle accelerator and detector technologies.

A large electron–positron (LEP) collider was constructed at the particle accelerator CERN and operated at energies up to 209 GeV. Four different experimental groups[5] hunted for the Higgs boson at this facility. The main focus was the associated production of the Higgs particle along with a Z boson. In this scheme, shown in Fig. 11.1, the high-energy electron–positron pair (e^-) annihilates, producing a virtual Z boson, also of high energy. This transforms into the Higgs

---

[4]John Ellis, Mary K. Gaillard, and D. V. Nanopoulos, "A phenomenological profile of the Higgs boson," *Nuclear Physics B*, **106** (1976): 292–340. While this is a technical paper not intended for non-particle physicists, it should be assigned as a *must-read* for any aspiring particle physicist, regardless of their current research focus.

[5]These experiments were denominated by the following acronyms: ALEPH (Apparatus for LEP pHysics); DELPHI (DEtector with Lepton, Photon, and Hadron Identification); L3 (Third Letter of Intent); and OPAL (Omni-Purpose Apparatus for LEP). These four initiatives are well-described in the book, *LEP—The Lord of the Collider Rings at CERN 1980–2000*, by Herwig Schopper (Berlin: Springer, 2009). Schopper was the Director General of CERN from 1981 to 1988.

boson (H) and a physical Z. The Higgs, in turn, becomes a bottom and antibottom quark pair, each of which appear in the apparatus as jets. The physical Z boson has its own recognizable modes of decay into quark jets.

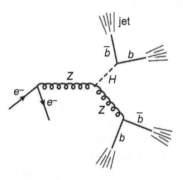

Figure 11.1 Production of the Higgs boson (H) in the LEP at CERN, beginning with a high-energy electron–positron annihilation. The electron is indicated by the leftmost e^- arrow; the positron traveling into the collision would be equivalent to an electron traveling backward in time, out of the point of collision, hence the second e^- arrow.

This work was carried out until the shutdown of the LEP at the end of the year 2000, to make way for the Large Hadron Collider (LHC). Ultimately, the LEP collaborations concluded that the Higgs particle must have a mass greater than 114.4 GeV/c^2.

At around the same time, two groups at Fermilab in the United States (CDF and DZero[6]) were also involved in a search for the Higgs using proton–antiproton beams of 1 TeV energy. These groups focused on Higgs production along with either W or Z bosons. This facility had the clear advantage of a large range of accessible energies, but high energy also results in a greater complexity of the background events that must be analyzed and sifted to find information that is relevant to one's search. The Fermilab groups claimed to have found the signal for a Higgs discovery on July 2, 2012, and they set a possible range of 115–140 GeV/c^2 for the particle's mass, specifying the most likely value around 125 GeV/c^2. But the final announcement of discovery and quantitative determination of the Higgs boson would finally belong to the LHC, announced by CERN at a global press conference two days later.

---

[6]CDF: Collider Detector at Fermilab; Dzero: Detector at the "D0" location of the Tevatron ring site.

Despite the large number of background events obscuring the physics of interest, proton colliders span a larger dynamic range of energies with their hadron beams than the lower energies available to the electron–positron colliders. The Fermilab search employed protons and antiprotons, but colliding beams that do not involve antimatter can have higher beam fluxes, since the production of antimatter involves a preliminary process of generating the particle–antiparticle pairs and then separating them from one another before finally accelerating them to high energies. Using only normal protons, the beam could be produced directly, by stripping hydrogen gas of its electrons so that high-intensity beams are available for the collisions. However, this will also lead to greater complexity in the resulting events, and thus more complicated analysis.

CERN, after evaluating the associated merits and complications, moved to a high-energy, high-intensity LHC proton–proton collider to expand its hunt for the Higgs boson and other interesting physics.[7] It is to the credit of the CERN groups that they realized most of the former LEP infrastructure could be converted into the new LHC at a reduced cost compared to the partially funded and later abandoned Superconducting Super Collider (SSC) of the United States.[8] During the 1990s, the Nobel Laureate and then CERN director-general Carlo Rubbia strongly promoted the LHC project around the world.

The LHC is not a single particle accelerator, but several working in tandem. Figure 11.2 shows a sketch of some of its components. Initially, protons are created at point P on the left of the diagram, accelerated to 750 keV energy by a pre-accelerator, and then to 50 MeV by a linear accelerator (LINAC). They are injected into a proton synchrotron booster (PSB) ring where their energy is increased to 1.4 GeV. These beams are further accelerated to 25 GeV in a proton synchrotron (PS), then receive an 18-fold increase in energy up to 450 GeV in a super proton synchrotron (SPS). At that point, the particles are fed into the main LHC ring to achieve the ultimate energy of 7 TeV, as two counter-propagating proton beams; their combined kinetic energy, in opposite directions, yields 14 TeV for the collision.

---

[7]The LHC is the latest in a series of accelerators built by the CERN groups, comprising collaborations from around the world. The timeline of CERN's history, highlighting its accomplishments, can be found at http://timeline.web.cern.ch/timelines/the-history-of-cern/overlay (accessed March 2018).

[8]A post-mortem of the SSC project may be found at https://www.scientificamerican.com/article/the-supercollider-that-never-was (accessed March, 2018).

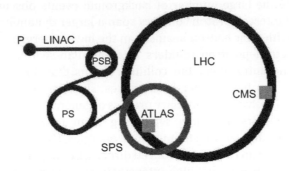

Figure 11.2 Large Hadron Collider (LHC) at CERN.

This scale of energy was a profound advance on earlier technology. As well, in 2014 the LHC machine had a sensitivity more than four times that of Fermilab's Tevatron, and since then the sensitivity has continued to improve. The advertised cost of CERN's machine, not including the contributions from external user countries, is over four billion Swiss francs, or approximately four billion US dollars.

After accelerating and colliding the proton beams, the LHC has two main detector systems that are employed in the Higgs boson search: ATLAS and CMS.

ATLAS is an acronym for "A Toroidal[9] LHC ApparatuS." The ATLAS detector is the size of a three-story building and has different layers of technology that are sensitive to different kinds of particles, including photons, electrons, positrons, muons, pions, protons, and neutrons. The detector layers are designed so that a particle will pass through each of them. Depending on what kind of particle it is, it may leave a track or it may not, and it may—in one of the layers—finally deposit its energy and end its life. The data on which layers register the particle and which do not, and where it finally comes to an end, all combine to provide an identification for the particle species.

CMS stands for "Compact Muon Solenoid." This detector system is a superconducting solenoid operating with a 3.8 tesla magnetic field. It consists of similar detectors as in ATLAS but differs in its geometry and materials, as well as its operational characteristics. These distinctions are intentional so that the two detectors are not

---

[9]A toroid is a doughnut shape, and the structure of ATLAS includes a doughnut-shaped magnet.

simply duplicates but ensure an independent study of the relevant physics. This protects any potential discoveries that might be made at the LHC from the criticism that they are mere instrumental artifacts.

Common to all high-energy experiments, the search for the Higgs boson at LHC began with a theoretical analysis. An enormous number of numerical simulations were performed using a computational technique called *Monte Carlo simulation*, which invests known or suspected physics with randomly generated data that is supposed to act like real events in Nature. Monte Carlo techniques were used to specify the parameters required to set up the equipment for best results, including beam characteristics and detector assembly. Also these techniques were applied to set criteria for the analysis of actual data expected from the experiment, and the methods that would be used to sift and analyze it. Over the last few decades, groups at CERN developed a unique set of software tools[10] accessible to their users from around the world, and to their great credit, these tools have been made available for general access to all non-profit academic users.

After several decades of work by countless researchers, theoretical numerical estimates for the production rate of the Higgs boson from proton–proton collisions were put in place to guide the design of LHC experiments. These estimates predict how many Higgs bosons should be expected to arise through different "channels" or processes that arise from collision energies between 7 and 14 TeV, the range available to the machine. Figure 11.3 illustrates[11] these numerical predictions, showing different channels for Higgs production (e.g., pp→H) along with a parenthetical explanation of the level of modeling involved in the theoretical calculation. The vertical axis is a logarithmic scale related to the number of events that are expected to produce a Higgs boson for a given energy of proton–proton collision, which is shown on the horizontal axis. "NLO" means

---

[10]"GEometry And Tracking," or GEANT, is a culmination of several decades of computer program development by CERN groups and their collaborators. Today's version, GEANT4, is highly sophisticated—a menu-driven user-code-specific object-oriented program. It has roots back to computer algorithms from the early 1960s. It might also be mentioned that the World Wide Web originated at CERN as a tool to serve the data-sharing needs of its globally distributed community of scientists and also for applied research in medical physics and related areas.

[11]C. Patrignani *et. al.* (Particle Data Group), "Status of Higgs boson physics," *Chin. Phys. C*, **40**, 100001 (2016), 11.

"next to leading order;" "NNLO" is "next to next to leading order;" and "NNLL" indicates "next to next to leading logarithmic." These refer to increasingly complicated calculations involving corrections to the most straightforward physical assumptions in the theory.

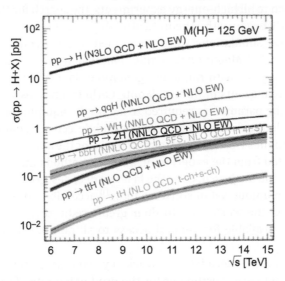

Figure 11.3 Model predictions for Higgs boson production from proton–proton collisions. The contributions of various individual channels of production and the total production cross section (top curve) are shown. Note the cross section is plotted in logarithmic scale (Figure courtesy: LHC HIGGS XS WG 2016).

The width of the lines in the figure indicates the amount of model uncertainties assumed to be present in the calculations for each channel. A narrower line, therefore, corresponds to greater confidence. These uncertainties may arise, for example, from those in the model's input parameters or imprecise estimates of effects due to other poorly understood phenomena.

As the collision energy is increased from 7 to 14 TeV, the rate of Higgs production is expected to increase by a factor of three. In the theory, most of this production is the result of gluon–gluon fusion, which transforms into a virtual top and antitop quark pair, which then becomes the Higgs boson. Lighter quarks are expected to contribute very little to the process due to the huge mass difference between the top quark (175 GeV/c^2) and the next heaviest bottom quark (about 5 GeV/c^2). In making the estimates, the theory assumes the limit of an infinite mass for the top quark, by matching the Standard Model to an effective theory.

Once the experiment is operational, we need to sift through the data from the detectors looking for specific events that bear the signature of a Higgs boson that has been born, briefly lived, and then died through decay into new entities. Figure 11.4 provides present-day model estimates[12] of the mass we can expect for a Higgs particle, in the range of 120 to 130 GeV/c^2, correlated with the various ways that the Higgs might decay. "ZZ," for example, corresponds to a decay mode where the Higgs boson transforms into a pair of Z bosons.

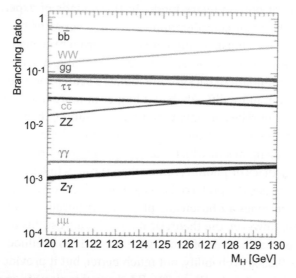

Figure 11.4 Theoretical decay modes of the Higgs boson (Figure courtesy: LHC HIGGS XS WG 2016).

The ZZ and WW modes (denoting W bosons) stand out in the plot with their increasing slope. They have a likelihood of 5 to 50 times that of decay by emission of two photons ($\gamma\gamma$). There is a curious detail in the physics of these two channels that requires some consideration. The mass of a Z boson is about 90 GeV/c^2 and that of a W is about 80 GeV/c^2. Thus, the combined mass of a pair of Z's or a pair of W's would exceed the mass range indicated on the horizontal scale of the graph, which is supposed to provide the theory's range of rest mass predictions for the Higgs that creates each pair, around

---

[12]Patrignani, 11.

125 GeV/c^2! The escape clause of Heisenberg uncertainty comes to the rescue here: In these decay channels, at least one of the W or Z bosons must be produced as a virtual particle, thus allowing a momentary violation of otherwise entrenched physical conservation laws.

The most prominent mode is the bottom quark pair, b$\overline{\text{b}}$. Research at LEP and Fermilab focused on identifying the Higgs through this decay channel, to no avail. The actual discovery of the Higgs, by the CERN groups, involved the two-photon ($\gamma\gamma$) channel, even though this mode is less likely by a factor of about 200 times compared with the dominant bottom quark mode. In these kinds of experimental searches, the main considerations are the amount of background produced (which must be sifted out), and a proper reconstruction of the mass of the particle for which we are searching. A decay channel with a large amount of background makes a search for a needle in a haystack even more difficult, by adding even more hay. If the mass reconstruction relies on data with poor resolution, then the needle is better camouflaged against the background of straw. Therefore, one looks for modes in which the decay will go more directly into charged particles or photons rather than through indirect steps.

In this sense, the preferred decay mode might be the one where the Higgs becomes a Z boson and photon (Zγ), because it offers good resolution. However, the theoretical models suggest this as close to the least probable mode of decay, with only 0.15% likelihood. The $\gamma\gamma$ mode has 0.23% probability, not much better, but it provides better mass resolution for the Higgs. The ZZ channel is also of comparable resolution to the two-photon process, but it comes with a larger background to sort through.

Thus, the $\gamma\gamma$ channel appears to provide a reasonable compromise for analysis. As in the case of experimental searches for the W and Z bosons, one ignores data from particles and photons that travel along the line of the collision of incident beam particles. Instead, the experiment looks for photons with large momentum in a direction perpendicular to the beam path, known as *high transverse momentum* photons.

The photons involved in the $\gamma\gamma$ channel have energies of several tens of GeV. At these energies (indeed, even above a few MeV), the inevitable interaction of gamma rays with a detector is to cause *showers* of high-energy charged particle pairs that, in turn, emit high-energy radiation, which causes further showers, in an escalating process. The two-photon signature discussed above is, therefore, an oversimplification of the complex pattern of the real event. These so-called showers extend in space, propagating as divergent cones from their point of origin, along the trajectory of the photon that causes them. It is practically impossible to capture the entirety of this effect and measure the full energy of the original photons using finite physical detector elements. Experimenters thus employ *sampling calorimeters*. When the photons encounter these elements of the detection apparatus, they alternately pass through an absorbing medium and a detecting medium. The shower cone is thus contained in a small volume until the energy content of the photon is reduced to only a few MeV, which is a level of energy that can be neglected in comparison to the very high energies of the original photons and secondary particles. The sample calorimetry measurements provide the energy and direction data for the photons, which we can use to deduce the relativistic invariant mass of the photons' parent particle, which is, perhaps, a Higgs boson.

11.3   Higgs Discovery

Figure 11.5 shows the results[13] announced by the ATLAS group as a confirmation of the Higgs boson discovery. The top portion of the graph shows the relativistic invariant mass of $\gamma\gamma$ photon pairs, selected from the full set of experimental data for high transverse momentum. The dashed curve, identified as "Bkg (4th order polynomial)," is the estimated background that needs to be cleaned away to reveal the presence of Higgs bosons in the data. The solid curve is the prediction of the Standard Model of particle physics for the joint contribution of the Higgs signal along with background noise.

The bottom part of the graph shows the final "peak" formed from a handful of possible Higgs boson appearances, after the background noise is subtracted away. The isolation of this peak is the final

---

[13]ATLAS Collaboration, "Plot of invariant mass distribution of diphoton candidates after all selections of the inclusive analysis for the combined 7 TeV and 8 TeV data," https://cds.cern.ch/record/1605822 (2013) (accessed April 2018).

consequence of immense dedication and ingenuity on the part of the particle physicists: It involved the sensitive identification of about 1000 events of physical interest from an overall data set involving trillions of collisions in the detector volume.

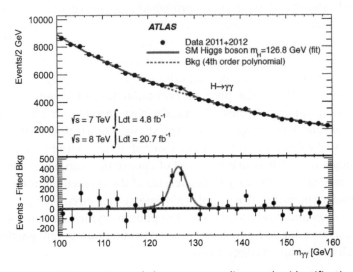

Figure 11.5 LHC experimental data corresponding to the identification of a Higgs boson. The top curve shows the 2-gamma events for center of mass energies between 100 and 160 GeV, selected from a few trillion collisions. The dashed curve, which overlaps with the solid curve everywhere except near 125 GeV, represents background events unrelated to Higgs production. The bottom figure shows the result of subtracting this background from the data: a remaining bump that provides evidence for the Higgs boson with a mass of 125 GeV/c^2.

The CMS experiment also found a Higgs boson signal comparable to that of ATLAS, in the $\gamma\gamma$ channel. As well, both CERN groups identified evidence for the Higgs decaying along the ZZ mode,[14] where, as mentioned above, at least one of the Z particles must be a virtual entity. While the masses, as determined separately in the two decay channels, do not completely agree, the average of the ATLAS values is compatible with the CMS result. This appears to provide satisfactory confirmation leading to a consensus of the

---

[14]The ATLAS collaboration announced in May 2017, "The new data allowed ATLAS to perform measurements of inclusive and differential cross sections using the 'golden' H $\rightarrow$ ZZ* $\rightarrow$ 4l decay" [https://atlas.cern/updates/physics-briefing/higgs-golden-channel (accessed March 2018)]. It should be noted that the detectors see four leptons from which the ZZ* and, subsequently, the Higgs are reconstructed.

particle physics community that the Higgs boson has been positively identified.

11.4   Particle or Resonance?

The year 2012 was hailed as the watershed moment of particle physics, with the announcement of the Higgs discovery at CERN.[15] The Higgs discovery was understood to be the missing puzzle piece in the Standard Model, although it is also argued that much work remains to be done in understanding the true rules and building blocks of the physical world. In light of the expense of time and resources involved in the search for the Higgs particle, the overwhelming public attention accorded to its discovery, and the degree of theoretical significance attributed to its experimental observation, it behooves us to reflect for a moment—at least—on a different way of thinking about what was observed in the CERN data set.

We should begin by summarizing what we know of the structure of interest, depicted in Fig. 11.5. It certainly has an energy of 126 GeV and is of zero charge, corresponding to the two-gamma final state of some intermediary entity. The question lingers as to whether the 126 GeV structure should be considered a *particle* or a *resonance*. Let us remind ourselves about the distinctions between these two-particle physics concepts.

First, a particle-like structure will have a reasonably long lifetime in terms of a nuclear physics timescale. It is generally understood (though not strictly adhered to) that a true particle lives longer than about a femtosecond (10^{-15} seconds). In the context of a particle physics experiment like the one in which the Higgs boson was identified, this kind of lifetime should correspond to a peak in a graph like the one in Fig. 11.5 with a width (called a *decay width*) spanning a narrow range of less than 1 keV. The theoretical model estimates for the Higgs' decay width are on the order of 4 GeV, translating to a lifetime less than 10^{-24} seconds. Thus, the much-publicized Higgs

---

[15]More precisely, the announcement occurred on July 4, 2012 (American Independence Day) at a well-attended event involving many high-profile physicists at CERN in Geneva, Switzerland. Perhaps fittingly, the momentous announcement date was chosen to coincide with the International Conference on High Energy Physics being hosted in Australia.

boson peak from 2012 dramatically fails this test to be considered a true particle.

Second, when we consider the creation of a particle in a collision experiment, the particle will only come into existence if a sufficient amount of energy is supplied to account for its mass, according to the rules of mass–energy equivalence in Einstein's special theory of relativity. If the energy supplied to the system in the collision is greater than this requirement, there is no reason (apart from certain prohibitions that arise from symmetry considerations) to suppose that the particle will not be produced in the laboratory. The extra energy left over after accounting for the particle's mass will simply be converted to kinetic energy in the new particle's motion. Indeed, this is what we see for objects that are undeniably recognized as subatomic particles such as pions, muons, and kaons: When the mass threshold is exceeded, these particles continue to appear in the detectors, with greater momentum. In searching for the Higgs boson, however, the experiments at Fermilab had adequate beam energies (at the TeV level) but gave negative results. If the Higgs were a true particle, we should expect that extra energy to become kinetic energy in the reaction products and still reconstruct the Higgs mass using special relativity, just as it was done at CERN. The Higgs boson candidate does not seem to naturally pass this excess energy test for particle status, either.

We have discussed in Chapter 5, on the topic of mass, that the particle physics community seems to follow arbitrary rules when deciding which entities to call a particle and which to subordinate to the status of resonance. We showed that the long-lived, narrow structures of charmonium and upsilonium were relegated to resonance status in order to reserve the desired ontological quality of "physical particle" for the charm and bottom quarks that are their constituents. However, the Z structure that appears with an energy of 90 GeV and decay width of 2.5 GeV is deemed a particle, even though its lifetime of 10^{-24} seconds. is comparable to, or shorter than, many processes on the nuclear scale that should otherwise render it only a temporary physical anomaly. The Z, conceptualized solely for its role in the neutral current *weak interaction*, decays by emitting *strongly* interacting hadrons, as understood within the quantum chromodynamics theory of *strong interactions*. This discrepancy is understood by quasirealists to represent the true physics of our world—to provide insight into the real structure of physical reality.

Should we not pause to ask if these quasirealist conclusions are supported by logical rigor, insofar as they are interpreted, believed, and communicated to say something substantial about reductionist physical reality? The same question, then, may be asked of the Higgs boson, which appears to arise in our data with all the qualities of a resonance but is slotted in to a particle role in our physical explanations.

11.5   What Does the Higgs Boson Contribute?

According to the current particle physics orthodoxy, particles that otherwise would have no mass acquire a finite mass property due to their interaction with the Higgs field, mediated by the Higgs boson. A particle's mass value is proportional to how strongly or weakly that body interacts with the Higgs, so if we know the mass of a particle, we can specify the amount of interaction. Conversely, if some model predicts the degree of interaction of a specific particle with the Higgs boson, we can then calculate the particle's mass. For example, since a proton is 1836 times as massive as the electron, we can say that protons couple 1836 times as strongly with the Higgs field than electrons do.

Our current theory, however, is silent as to *why* a particle should have the specific interaction strength that it does. The mass of a fermion is proportional to the coupling; the mass of a boson is proportional to the square of the coupling; and the mass of the Higgs boson, itself, is due to self-coupling. In this picture, gluons and photons do not interact with the Higgs in the most straightforward calculations referred to as *zeroth order*. However, higher-order calculations that represent increasingly unlikely activity of virtual particles, including creation of particle–antiparticle pairs, do introduce a coupling between the Higgs boson and gluons and photons. These couplings will induce mass to virtual versions of these physically massless particles.

We should not forget that all these couplings and interactions occur in a complex *isospin space*, a mathematical space that is defined to aid our calculations by analogy to real, physical space. We refer the reader again to Chapter 4 of this book, dealing with mathematical spaces, where we discuss the usefulness of these concepts, but also the associated danger of quasirealist interpretations. In the case of our theory of the Higgs field, we can invoke ideas that follow by mathematical analogy to superconductivity in solid-state physics, as prescribed by Nambu[16] and others, but we are dealing with a mathematical space and not with the actual physical universe and its direct properties. Some of the interpretations of the way the Higgs field and its boson operate in our world may be no different from conceptualizing a complex impedance in the context of real capacitors, inductances, and resistors in a real electrical circuit. Concepts like these are useful but should not be considered representations of real physical entities. To argue that the Higgs boson is a physical particle related to a physical field permeating the whole universe, responsible for the physical property of mass of other (more demonstrably) physical particles, is quasirealist overreach.

It is also opportune at this moment to reflect on the role of mass in particle physics. It does not play the same role here as it does in the context of macroscopic physics, where mass is a measure of the inertial response of a body to an applied force, whether that be of a gravitational, mechanical, or electromagnetic variety. Under those conditions, macroscopic mass is a well-defined constant property of a specific object with which we characterize its response to external conditions. We can define the density of a body as its mass per unit volume and estimate how the behavior will change if we increase the material content of the body. In special relativity, mass lost its distinct position in the description of material dynamics and was redefined as another form taken by energy, its magnitude dependent on a body's relative motion with respect to a measuring instrument. In general relativity, the situation becomes more obscure. Forces are absent in the general theory, replaced by spacetime curvatures. It was only in the second half of the twentieth century that physicists defined Komar and Bondi mass with respect to the definition of geodesics.

In microscopic physics, we use the concept of mass in both ways already mentioned: as a form of energy and as a measure of inertia. Chemists as well as atomic, molecular, and nuclear physicists

---

[16]Yoichiro Nambu (1921–2015) was a Japanese-born American physicist. He was awarded the Nobel Prize for physics in 2008 for work on spontaneous symmetry breaking in subatomic physics. His theory, based on an analogy with superconductivity, inspired the Higgs mechanism.

use mass to assess the energy of a system, for example, to evaluate the input energy required for a certain reaction or to estimate the energy released in various processes. These concepts are essential to practical, industrial applications like the energy release in nuclear fission for power generation. The mass concept as a measure of inertia is used to determine particle trajectories in the presence of external electromagnetic fields.

In particle physics, the equivalence of mass and inertia is only used for those entities that leave their trails in physical detectors and is mainly relevant for protons, pions, kaons, electrons, and muons. For everything else, mass is determined by the relativistic invariant mass concept involving energy and momentum measurements. When we begin dealing with virtual particles, which are essential features of all theoretical treatments (involving, for example, higher-order Feynman loop diagrams), the particles are said to be "off mass shell," which means that they are not considered to possess their usual, nominal mass. For example, a physical W boson is understood to have a mass of 81 GeV/c^2. If we consider a W boson exchange process in the decay of a neutron, the energy available is less than 1 MeV for the W, which is clearly too small to account for the nominal mass of this particle's appearance. The theory accommodates this apparent impossibility by invoking the guiding hand of the Heisenberg uncertainty principle to generate just the required amount of energy, out of nothing, with the stipulation that the boson lives for less than about 10^{-25} seconds, thus traveling much less than 10^{-17} meters before it dies.

Both relativity and quantum field theory stipulate that particles of zero mass cannot have longitudinal polarization. When attempts at electroweak unification stubbornly yielded four massless bosons, we introduced a complex scalar field in (mathematical) isospin space to induce longitudinal polarization for three of them, thus introducing three massive bosons. It is often, perhaps lightheartedly, remarked that the Higgs field "ate" the three massless bosons. However, as discussed in Chapter 8 dealing with the concept of elementary quanta, none of the three physical weak bosons, nor photons, is a pure, well-defined physical state; as understood by our latest theory, they are composites of mathematical entities. The physical reality of all this—as opposed to its mathematical cleverness and undeniable utility—remains obscure.

As of now, it is the Higgs field and boson that are said to give mass to other particles. However, some models suggest that the interaction of the Higgs with other, heavier particles may eat up the Higgs just as the Higgs field ate up the three massless bosons. This problem is called the *naturalness* or *hierarchy problem*. It leaves us with a picture of the Higgs' world of particle physics as a ferocious marine food chain with big fish eating small fish, only to be devoured—perhaps—by sharks or whales occupying places higher up in the increasingly complicated subatomic zoo. This Higgs boson may be just one entity among many other predatory bosons waiting to be discovered. Particle physics often describes their discipline as a study of the "subatomic zoo." It may be more appropriate to use the analogy of a vicious marine ecosystem.[17]

11.6   Conclusion

The prediction and discovery of the Higgs boson has been, unarguably, an exciting chapter in the history of modern physics, entailing countless hours of ingenious theoretical and technical problem-solving on the part of thousands of physicists and mathematicians over several generations. Let us recapitulate what we can say of the 2012 discovery.

The structure at 126 GeV, while it appears as a broad resonance (with a width of about 4 GeV, which is nearly equal to the mass of an alpha particle or helium atom), is commonly assigned the ontological status of a particle. It was identified in the two-photon decay channel, despite the curious reality that photons are understood to be of zero mass due to their non-interaction with the Higgs. A structure corresponding to the Higgs boson is also seen in a decay channel where it becomes a ZZ* pair involving a virtual Z*, and they subsequently decay into four leptons. Oddly, again, these leptons are particles of very small mass and thus small coupling to the Higgs field.

---

[17]Learning that Nobel Laureate Leon Lederman (b. 1922) seemingly chose to call the Higgs boson the "God particle," one of us was inspired to give a conference presentation with the title, "Higgs Boson—God Particle or Divine Comedy?": C. Rangacharyulu, *Proceedings Volume 8832, The Nature of Light: What are Photons? V*; 88321C (2013); doi: 10.1117/12.2027833.

Remembering that, according to current particle physics interpretations, all gauge bosons and many intermediary interactants live in virtual states, which may be far removed from the nominal masses ultimately assigned to them, we may question whether Higgs physics contributes any essential clarity to the ancient quest of physical reductionism. Perhaps it has, but certainty on this point is obscured by the intimate melding of theory with the interpretation of data from the Higgs experiments, and ambiguity surrounding the definition of mass—the very property that the Higgs field is supposed to impart to matter. We are concerned that quasirealist certainty about the ontology of the Higgs particle and field may have, in fact, led the particle physics community away from the reductionist vision of the discipline and into a more convoluted understanding of Nature than is strictly necessary. This may do a disservice to the real effort and expense that has accompanied the exceptional theoretical work related to investigations in the *mathematical* world of quantum field theory.

Epilogue

I also remember a remark of Albert Einstein, which certainly applies to music. He said, in effect, that everything should be as simple as it can be, but not simpler.

—Roger Sessions[1]

Physicists should be jubilant. One of the main dreams of our discipline may finally be realized. If so, the credit will go to the hard work of the nuclear and particle physicists. At least, we see indications of this dream fulfilled in the messiness that characterizes our current theories and definitions at the bottom of physical reality.

The history of physics has included a grand, sweeping commitment to the idea of physical reductionism, the philosophy that all the stuff of Nature finds ultimate causal explanation at lower and lower levels of physical scale. The aim of every discipline of science is to work out methods to describe relevant complex systems in terms of simple concepts directly related to observable structures, identifiable behaviors, and measurable responses to external conditions. Such methods almost always incorporate reductionism in some form, and this is true whether the sciences are biological, social, or physical. The breadth of reductionism applies to the study of whole societies, individual human and animal psychology, the occurrence of geological events, and the interactions between the submicroscopic parts of chemical atoms. At the bottom of the hierarchy of reductionist explanations we find the particle physicists hunting for the final rung of the ladder. The philosophy of physical reductionism insists that whatever is found at the bottom

---

[1]"How a 'Difficult Composer Gets That Way," *New York Times* (New York: New York Times Company, January 8, 1950), 59.

will provide adequate causal explanation for all that lies above in the various macroscopic scales of matter.

We believe there is good reason, in our day, to suspect that particle physicists had stepped onto that bottom rung, completing the task of the reductionist quest. It might be time to celebrate this achievement. This book has been about clues that we see pointing to this conclusion.

First, however, this brings to mind an important caution. It is the subtle arrogance of many physicists to assume that *all* phenomena in the range of human experience can be described, in principle at least, by the rules that reduce to particle physics: a blind faith in the universal power of particles in motion. This is philosophical overreach, since no scientific evidence can verify whether or not reality is *only* composed of physical entities; but to some who work in this business it is somehow inconceivable that reality may include aspects that are ontologically distinct from physicality. For them, to accept any essential limitation on the power of physics would amount to a heretical denial of both the physical and theoretical reductionist ideology; an insufferable admission that Nature contains mysteries beyond our ability to systematize.

In reality, however, there appear to be some essential limits to the explanatory power of the physical sciences, the most obvious one being the study of individual human behavior. To reduce the subjective psychology and the experiences of an individual person to some lower level of physical causation we must first confront the reality that each human action is, we may say, connected to one or more first-person mental events. What triggers a mental event and how does an individual respond to it, retaining all the while their subjective impressions of control and qualitative experience, which make us ontologically distinct from hypothetical "zombies"?

If we insist, in a militant way, on applying physicalist explanations to mental events, we must admit that this will be a very difficult problem to solve from a reductionist perspective. No two people respond in the same way to the same event and even a single individual might react very differently on different occasions. Admitting the central, organizing role played by the human brain in acquiring and sorting sensory data, and facilitating responses throughout the systems of the body, our reductionist explanations should be focused on this organ. We will naturally identify the brain as the seat of physical responsibility for human behavior. It is a simple matter to locate it, hidden in the skull, and estimate its size and other relevant properties; it is a more difficult, but not impossible, task to deconstruct the human brain down to its microscopic construction and describe the intricate flow of chemicals throughout its biological mass.

At this point, however, we have accomplished all we can. This is the nearest we get to attaching physical reality to human mental perceptions. Medical science informs us that the brain is one of the largest and certainly the most complex of the body's organs. Any attempt at a realistic computational model of the brain's electrochemical activity, beyond the barest similarities and most extraordinary simplifications, is beyond our wildest hopes, despite the breathless anticipation of those whose careers depend on a continued flow of research financing targeted toward the cartoon future of truly self-aware artificial intelligence.[2] Algorithms are nothing but lines of code mechanically transformed into the motion of electrons in conducting solids. No C++ procedure, even when it is woven through the circuitry of multiple parallel processors, can ever result in a first-person mental perception. If it could (which implies that we have some inconceivable way of externally recognizing that such an event has taken place!), we should be able to highlight the exact line of code that makes the difference between inert physicality and the advent of magical consciousness. A misplaced pixelated semi-colon could then make all the difference for the existence of a new conscious being in our universe. There must be something more than that to consciousness.

Although mental events are a kind of upper limit for physicalism and reductionism, we can confidently apply our reductionist philosophy in the direction of lower levels. We identify biological cells as the next simpler entity out of which the physical brain is built. In principle, it is the collective interactions of cells that define the mechanical basis of the brain, and all the parts of the body. A

---

[2]This is not to say that future computer programmers would not produce artificial intelligence code that can pass the Turing test and trick an observer into believing that self-awareness lies beneath. This, however, says everything about the gullibility of the observer and nothing at all about the reality of true mental events in the artificial entity.

human body is said to consist of a few trillion cells. Imagine trying to reconstruct the effects of trillions of parts interacting (in some poorly understood way) with each other to make life do what it does. Recognize, as well, that these mutual interactions affect the properties of the entities, themselves, and thus recursively change the way the entities interact. The task of reducing this trillion-fold collective activity to the level of individual parts presents a nightmare in self-referencing complexity. Nonetheless, we can be philosophically confident, as reductionists, that our lack of ability does not change the underlying *reality* that a physical brain or body is nothing but a collection of cells. We can build models of that, however approximate.

So reductionism takes us to the next level down, where our confidence increases. We may ask, what are the cells made of? This is where we come to inanimate matter, devoid of whatever we identify as "life." At this stage, our field of vision is increasingly filled with nearly identical structural units of complex and simple molecules, electrical charges, and currents as the means by which cells hold together and interact. What they communicate and how they react to stimuli are still, certainly, a difficult and detailed problem, but not impossible to work out in principle. Things are becoming simpler. It is helpful that larger examples of these material entities can be seen and manipulated in the laboratory providing direct guidance in our search for patterns in their behavior.

Another step deeper takes us to chemical atoms. It is a remarkable achievement of our reductionist method that at this point, about 100 chemical elements in the periodic table of chemistry can be described by various configurations of three basic quanta: the world of electrons, protons, and neutrons, along with energy they exchange. We are in a comfortable position of being able to manipulate these entities using electromagnetic fields, and we can employ our classical notions of mass, weight, motion, *et cetera*, and see them flying in the laboratory with sensors that respond to the energy deposits and ionizations they cause. We can determine the time of the day at which a particle went in a specific direction and what its kinematical properties were at that moment, such as its momentum and energy. Unlike a charged particle, a neutron or a photon does not leave trails as it moves along, but it still does make its presence known as it interacts in a detecting medium. In that sense, the physical reality of all these entities can be considered established beyond any doubt.

From the philosophical perspective of reductionism, we have a causal chain linking this atomic scale of the world all the way back up to the physical brain. We started at the top of physical reality and explained each layer in terms of a lower set of mechanisms and entities until we arrived at the nucleons and electrons at the bottom. Along the way, we only ever met tangible structures, accessible to our tools in the laboratory. The proton, neutron, and electron, exchanging and emitting energy in the form of radiation, are responsible for all the dynamics that we see above the atomic level. As for the nature of the radiation, it can be discussed forever, but if nothing else we recognize that it is transitory, transporting energy from one material body to another and not a permanent component of matter. Thus, we have all our pieces in place and they add up to three distinct varieties.

Is this state of simplicity the end, then? As we have previously explained in the preceding chapters, Democritus' atomic theory specifies one simple criterion to evaluate that question. We will know the basic atoms by their indivisibility—not their infinitesimal size or lack of internal properties, but the fact that, try as we might, we cannot pull them apart into smaller building blocks. They are the *uncuttables*. The level of electrons and nucleons appears to meet this criterion and offer the extraordinary kind of simplicity (three building blocks!) required to finally celebrate the fulfillment of the vision of physical reductionism. In fact, we have attempted to probe to deeper levels and found things become immediately *more complicated* rather than less. Everything should be as simple as can be, but not simpler, as Einstein is said to have said.

The nuclear and particle physicists have not been listening to this advice, it appears. As they probed the physics of the smallest scale, their analysis suffered from a critical, practical limitation that led to an inappropriate interpretation that they were seeing a yet-deeper level of structure. The theoretical description of atomic nuclei and complex molecules was modeled on mutual interactions occurring only between two individual entities at a time, popularly known as two-body interactions. It is, of course, impossible to imagine that in the medium of many real bodies, every pair of entities interacts with one other as if the others have no influence on that transaction.

Isaac Newton struggled with this issue as he worked out the earth–moon–sun dynamics as three pairs of two-body interactions and wondered if the contribution of a three-body force should be considered, where all three affect each other simultaneously. This three-body interaction, however, was not easily calculated, and this remains the case for atomic and molecular physicists. In that context, it is even more acute, because any real collection of matter involves not just three bodies interacting but some innumerable quantity of electrons and nucleons all contributing together to the structures and dynamics of the real system.

To make headway with this, physicists devised *mean field descriptions* in which the behavior of the majority of the particles in the system is approximated as a kind of background average potential influencing only a few specific bodies—perhaps one or two electrons—which provide generalized information about the system from their idealized response to the collective smear of everything else. Naturally, this technique yields information only about quasi-entities.

Sometime in the 1940s, nuclear physicists split into two groups: those who deal with complex nuclei, and the particle physicists who were intent on understanding the interactions between pairs of protons, neutrons, and electrons. The particle physicists were following the arrow of reductionism, attempting to account for the complex nuclei from the bottom up, considering all conceivable kinematical conditions and interactions among their parts. The former group were subject to criticism that their physics did not include enough fundamental science, and the latter group considered themselves to be addressing this problem head-on.

As these investigations probed matter at its smallest scales, the whole enterprise of particle physics received a rude shock. Experiments were being carried out at increasing energies, under various kinematical conditions, to understand the details of the calculable two-body dynamics. As the energies increased beyond a certain threshold, however, the interactions became dominated by the production of a third particle: pions. These supposedly two-body interactions were suddenly anything but two-body! With increasing energies, the complexity and involvement of extra bodies just became worse and worse. The subatomic zoo grew to include many more entities and the scientists forgot their initial goal of understanding nuclear dynamics.

Since then, understanding the subatomic zoo has, itself, become the main preoccupation of particle physics. It is a step sideways from the quest of physical reductionism, since none of these new particles that appear in high-energy collisions comprise building blocks of matter. They add nothing to the foundation of the proton, neutron, and electron.

As we have seen in the preceding chapters of this book, these efforts to say something useful about the subatomic zoo have instead confused and obscured our reductionism, trading the physics of real things for increasingly abstract mathematical concepts masquerading as structures in a quasireality. In the search to simplify Nature, to account for small discrepancies from theory in the data, particle physicists have produced a catalog of entities. On the one hand, these are presented to the public using classical, physical language; on the other hand, they are formally defined in theory as composites of mathematical structures of various flavors and colors, living in multidimensional mathematical spaces.

This activity has entailed a lot of mathematical modeling, based on the foundation of quasirealism in Einstein's relativity and the quantum theory. Particle physicists drew inspiration from their solid-state colleagues, experts at working with effective fields and quasiparticles. Indeed, the Nobel Laureate Yoichiro Nambu credited his success in particle physics to his early training in the field of condensed matter. This is evident from his research publications and Nobel lecture, which reveal the substantial influence of the BCS superconductivity theory on his thinking. As we have previously argued, quasiparticles are mathematical entities that compactify theoretical formulation but have no real, ontological status. A physical world described entirely by effective fields, mathematical spaces, and quasiparticles can only be a quasireality.

Figure E.1 provides a qualitative description of this result, sketching the relationship between the complexity required to describe a natural system, and the reductionist arrow of physical causation. It illustrates the two boundaries of complexity that define the limits of physical reductionism's descriptive power. At the left end of the figure, corresponding to the most complex physical structures in Nature, mental events are not physically accessible and cannot be modeled by physicochemical, mathematical, or computational theories. Yet we know mental events happen. They are real but not physical. It is a realm of mystery.

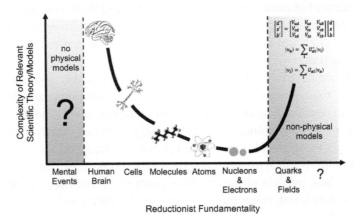

Figure E.1 The hierarchy of physical reductionism compared to the complexity of corresponding scientific models. The trend toward simplification appears to be violated when we reach the sub-nuclear level of quarks, neutrinos, and quantum fields. This may be a clue that we have reached the bottom.

Moving across the figure, *physical* reality begins with the physical brain and continues until we reach the proton, neutron, and electron as the smallest entities of which we can have clear ideas. The resources required to model the physical phenomena in between these levels steadily decrease, and we feel comfortable that there is a simple correspondence of mechanical cause and effect from one layer of complexity to another: nothing involved but some complicated interactions among physically tangible entities.

Arriving at the level of Nature represented on the right side of the figure, our reductionist quest encounters a problem. Further efforts to theoretically describe physical systems below the level of the proton, neutron, and electron cause an abrupt deviation from the trend toward simplification. Suddenly, our models are more complicated, increasingly disconnected from common-sense experience, increasingly reliant on abstract mathematics and effective methods, and less understood even by the practitioners of the field. The net result is anything but conceptual clarity, unless it is enforced through artificial simplifications that make use of the classical physical language that is conceptually accurate at the higher levels. Is it that we have attempted to make Nature simpler than it can be? We have entered another realm of mystery.

Thus, we think Nature may have an important lesson for us in these developments. Is it possible that all signs point to the bottom of Nature at the level of atomic parts? The arrow of physical reductionism should always point to simpler and simpler levels of description, just what we do not see in the details of the most recent particle physics discoveries. Our clue, then, is the abrupt reversal in the trend toward overall simplification when we probed the sub-nucleon level. Our best efforts to ground human mental events at the top of a physical substratum are a non-starter. So also—by symmetry, perhaps—the theoretical abstraction required in the modern analysis of quarks and gauge bosons and the Higgs field indicates the bottom.

If this suspicion is true, then it would amount to a most exciting conclusion: We have established the identity of the Democritian building blocks and they are protons, neutrons, and electrons. Perhaps it is time to celebrate this ultimate achievement of the twentieth-century particle physics, the fulfillment of the ancient reductionist dream.

Appendix

Epitaph for All Photons[1]

Oh Photon, thou art a bundle of energy, not a bundle of joy!
Thou causeth pain and agony to many a miserable soul
Who spendeth entire life in quest to figure thee out

Thou are the toughest nut to crack
But, when we crack thee thou art not there
Long gone or you were just dead

We also hear that thou findeth thyself in at least two forms
Real or virtual: Maybe we got virtual reality from you

Thou art the Phoenix, finding the sun in encounter with any thing
Thus comes a promise of eternal life to an abrupt end
At times setting off another photon on an indeterminate voyage.

—**Chary Rangacharyulu**

---

[1]Originally presented in: "An Epitaph for All Photons: A Phoenix Rising from Its Ashes," *Proceedings of SPIE*, vol. 9570, The Nature of Light: What Are Photons? VI, 95700I (2015), doi: 10.1117/12.2185705.

Index

Printed and bound by CPI Group (UK) Ltd, Croydon, CR0 4YY

23/10/2024

01777698-0002